Lecture Notes in Mathematics

Edited by A. Dold and B. Eckmann

486

Şerban Strătilă
Dan Voiculescu

Representations of AF-Algebras and of the Group U (∞)

Springer-Verlag
Berlin · Heidelberg · New York 1975

Authors
Dr. Şerban-Valentin Strătilă
Dr. Dan-Virgil Voiculescu
Academie de la Republique
Socialiste de Roumanie
Institut de Mathématique
Calea Grivitei 21
Bucuresti 12
Roumania

Library of Congress Cataloging in Publication Data

Stratila, Serban-Valentin, 1943-
Representations of AF-algebras and of the group
U( )

(Lecture notes in mathematics ; 486)
Bibliography: p.
Includes indexes.
1. Operator algebras. 2. Representations of algebras. 3. Locally compact groups. 4. Representations of groups. I. Voiculescu, Dan-Virgil, 1949- joint author. II. Title. III. Series: Lecture notes in mathematics (Berlin); 486.
QA3.L28 no. 486 [QA326] 510'.8s [512'.55] 75-26896

AMS Subject Classifications (1970): 22D10, 22D25, 46L05, 46L10

ISBN 3-540-07403-1 Springer-Verlag Berlin · Heidelberg · New York
ISBN 0-387-07403-1 Springer-Verlag New York · Heidelberg · Berlin

Offsetdruck: Julius Beltz, Hemsbach/Bergstr.

INTRODUCTION

Unitary representations of the group of all unitary operators on an infinite dimensional Hilbert space endowed with the strong-operator topology have been studied by I.E.Segal ([30]) in connection with quantum physics . In [21] A.A.Kirillov classified all irreducible unitary representations of the group of those unitary operators which are congruent to the identity operator modulo compact operators , endowed with the norm-topology . Also , in [21] the representation problem for the unitary group $U(\infty)$, together with the assertion that $U(\infty)$ is not a type I group , is mentioned .

The group $U(\infty)$, well known to topologists , is in a certain sense a smallest infinite dimensional unitary group , being for instance a dense subgroup of the "classical" Banach-Lie groups of unitary operators associated to the Schatten - von Neumann classes of compact operators ([18]) . Also , the restriction of representations from $U(n+1)$ to $U(n)$ has several nice features which make the study of the representations of $U(\infty)$ somewhat easier than that of the analogous groups $SU(\infty)$, $O(\infty)$, $SO(\infty)$, $Sp(\infty)$.

The study of factor representations of the non locally compact group $U(\infty)$ required some associated C^*- algebra . The C^*- algebra we associated to a direct limit of compact separable groups , $G = \varinjlim G_n$, has the property that its factor representations correspond either to factor representations of G_∞, or to factor representations of some G_n and, since the distinction is easy between these two classes, it is of effective use.

This C^*- algebra is an AF - algebra, i.e. a direct limit of finite-dimensional C^*- subalgebras. AF - algebras, introduced and studied by O.Bratteli ([1]), are a generalization of UHF - algebras. For the UHF - algebra of the canonical anticommutation relations of mathematical physics there is the general method of L.Garding and A.Wightman ([12]) for studying factor representations and, in particular, the cross-product construction which yields factor representations in standard form. So we had to give an extension of this method to AF - algebras (Chapter I). For $U(\infty)$ this amounts to a certain desintegration of the representations with respect to a commutative C^*- algebra, the spectrum of which is an infinite analog of the set of indices for the Gelfand - Zeitlin basis ([37]). For $U(\infty)$ in this frame-work, a complete classification of the primitive ideals of the associated C^*- algebra, in terms of a upper signature and a lower signature, is possible (Chapter III). Simple examples of irreducible representations for each primitive ideal are the direct limits of irreducible representations of the $U(n)$'s, but there are many other irreducible representations.

Using the methods of Chapter I, we study (Chapter IV) a certain class of factor representations of $U(\infty)$ which restricted to the $U(n)$'s contain only irreducible representations in antisymmetric tensors . This yields in particular an infinity of non-equivalent type III factor representations , the modular group in the sense of Tomita's theory ([32]) with respect to a certain cyclic and separating vector having a natural group interpretation. Analogous results are to be expected for other types of tensors .

The study of certain infinite tensor products (Chapter V) gives rise to a class of type II_∞ factor representations . As in the classical theory for $U(n)$, the commutant is generated by a representation of a permutation group . In fact it is the regular representation of the infinite prmutation group $S(\infty)$ which generates the hyperfinite type II_1 factor . Other examples of type II_∞ factor representations are given in § 2 of Chapter V

Type II_1 factor representations of $U(\infty)$ were studied in ([34],[35]) and the results of the present work were announced in ([38]) .

Concluding , from the point of view of this approach , the representation problem for $U(\infty)$ seems to be of the same kind as that of the infinite anticommutation relations , though "combinatorially" more complicated . Of course , a more group - theoretical approach to the representations of $U(\infty)$ would be of much interest .

Thanks are due to our colleague Dr. H.Moscovici for drawing our attention on [21] and for useful discussions .

The authors would like to express their gratitude to Mrs.

Sanda Strătilă for her kind help in typing the manuscript .

*

* *

The group $U(\infty)$ is the direct limit of the unitary groups $U(1) \subset U(2) \subset \ldots \subset U(n) \subset \ldots$, endowed with the direct limit topology . Let H be a complex separable Hilbert space and $\{e_n\}$ an orthonormal basis . Then $U(\infty)$ can be realized as the group of unitary operators V on H such that $Ve_n = e_n$ excepting only a finite number of indices n . Similarly , we consider $GL(\infty)$ the direct limit of the $GL(n)$ ' s .

By $U_1(\infty)$ we denote the group of unitaries V on H such that $V - I$ be nuclear , endowed with the topology derived from the metric $d(V',V") = Tr(|V' - V"|)$. Also , by $U(H)$ and $GL(H)$ we denote all unitary , respectively all invertible , operators on the Hilbert space H .

As usual , wo - topology means weak-operator topology and so - topology means strong-operator topology.

Since it might be useful for the reader to have at hand certain classical facts concerning the irreducible representations of $U(n)$, especially in view of Chapters IV and V, there is an Appendix about these representations.

The bibliography listed at the end contains, besides references to works directly used, also references to works we felt related to our subject. We apologize for possible omissions.

Bucharest, March 12th 1975. The Authors.

CONTENTS

CHAPTER I ON THE STRUCTURE OF AF - ALGEBRAS AND THEIR REPRESENTATIONS

The uniformly hyperfinite C^*- algebras (UHF - algebras) , which appeared in connection with some problems of theoretical physics , were extensively studied , important results concerning their structure and their representations being obtained by J. Glimm ([13]) and R. Powers ([24]) . They are a particular case of approximately finite dimensional C^*- algebras (AF - algebras) considered by O.Bratteli ([1]) , who also extended to this more general situation some of the results of J. Glimm and R. Powers .

Our approach to the representation problem of the unitary group $U(\infty)$ required some other developments , also well known for the UHF - algebra of canonical anticommutation relations . Chapter I is an exposition of the results we have obtained in this direction , treated in the general context of AF - algebras.

We shall use the books of J. Dixmier ([6],[7]) as references for the concepts and results of operator algebras .

If M_1 , M_2 , ... are subsets of the C^*- algebra A , then we shall denote by

$$\langle M_1 , M_2 , \dots \rangle \quad \text{or} \quad \langle \bigcup_n M_n \rangle$$

the smallest C^*- subalgebra of A containing $\bigcup_n M_n$ and by

l.m.(M_1 , M_2 , ...) (resp. c.l.m.(M_1 , M_2 , ...))

the linear manifold (resp. the closed linear manifold) spanned by $\bigcup_n M_n$. Also , for any subset M of A , we shall denote by M' the commutant of M in A :

$$M' = \{ x \in A \; ; \; xy = yx \; (\forall) \; y \in M \} .$$

A maximal abelian subalgebra (abreviated m.a.s.a.) of a C^*- algebra A is an abelian C^*- subalgebra C of A such that $C' = C$.

A conditional expectation of a C^*- algebra A with respect to a C^*- subalgebra B in A is a linear mapping $P : A \longrightarrow B$ such that :

1) $P(1) = 1$;

2) $\|P(x)\| \leqslant \|x\|$ for all $x \in A$;

3) $P(x) \geqslant 0$ for all $x \in A$, $x \geqslant 0$;

4) $P(x)^* P(x) \leqslant P(x^*x)$ for all $x \in A$;

5) $P(yxz) = yP(x)z$ for all $x \in A$, $y,z \in B$.

Obviously , a conditional expectation of A with respect to B is a (linear) projection of norm one of A onto B . Conversely , J. Tomiyama ([33]) has proved that any projection of norm one of A onto B is a conditional expectation . In what follows we shall use the result of J. Tomiyama only in order to avoid some rather tedious verifications .

An approximately finite dimensional C^*- algebra (abreviated AF - algebra) is a C^*- algebra A such that there exists an ascending sequence $\{A_n\}_{n \geqslant 0}$ of finite dimensional C^*- subalgebras in A with

$$A = \langle \bigcup_{n=0}^{\infty} A_n \rangle \quad (= \overline{\bigcup_{n=0}^{\infty} A_n}) \quad .$$

We shall suppose that A_0 is one dimensional , $A_0 = \mathbf{C}\cdot\mathbf{1}$, where $\mathbf{1}$ stands for the identity element of A .

For C^*- algebras A and B , $A \simeq B$ will denote some obvious (star) isomorphism , in which case corresponding elements will sometimes be denoted by the same symbol .

§ 1 Diagonalization of AF - algebras

Given an arbitrary AF - algebra

$$A = \langle \bigcup_{n=0}^{\infty} A_n \rangle \quad ,$$

we shall construct a m.a.s.a. C in A , a conditional expectation P of A with respect to C and a group U of unitary elements of A , related to a suitable " system of matrix units for the diagonalization of A with respect to C " , such that

$$A = \text{c.l.m.}(UC) \quad .$$

I.1.1. We define by induction an ascending sequence $\{C_n\}$ of abelian C^*- subalgebras in A :

$$C_0 = A_0 \quad ; \quad C_{n+1} = \langle C_n , D_{n+1} \rangle \quad , \quad n \geqslant 0 \quad ,$$

where D_{n+1} is an arbitrary m.a.s.a. in $A_n' \cap A_{n+1}$.

LEMMA . <u>For all</u> $n \geqslant 0$ <u>and all</u> $k \geqslant 0$ <u>we have</u>

(i) C_n <u>is a m.a.s.a. in</u> A_n ;

(ii) $A_n' \cap C_{n+k}$ <u>is a m.a.s.a. in</u> $A_n' \cap A_{n+k}$;

(iii) $C_{n+k} = \langle C_n , A_n' \cap C_{n+k} \rangle$.

<u>Proof</u> . (i) The claim is obvious for $n = 0$ so we suppose it is true for C_n and we prove it for C_{n+1} .

Consider $x \in A_{n+1} \cap C_{n+1}'$. If p is a minimal central projection of A_{n+1} , then $p \in D_{n+1} \subset C_{n+1}$, pA_{n+1} is a factor and we have :

a) pC_n is a m.a.s.a. in pA_n .

Indeed , since the map $y \longmapsto py$ is a $*$ - homomorphism of A_n onto pA_n , there is a central projection z in A_n such that $pz = p$ and such that the above map is an isomorphism of zA_n onto pA_n . If $y \in A_n$ and if py commutes with pC_n , then $zy \in A_n$ commutes with C_n , thus $zy \in C_n$, since C_n is a m.a.s.a. in A_n . It follows that $py = p(zy) \in pC_n$.

b) pD_{n+1} is a m.a.s.a. in $(pA_n)' \cap (pA_{n+1})$.

This is clear , since p belongs to the center of $A_n' \cap A_{n+1}$.

c) $px \in pA_{n+1}$ commutes with $pC_{n+1} = \langle pC_n , pD_{n+1} \rangle$.

This is obvious .

If from a) , b) , c) we infer that $px \in pC_{n+1}$, then we have proved that $px \in C_{n+1}$ for any minimal central projection p of A_{n+1} . Since 1 is a finite sum of minimal central projections of A_{n+1} , it follows that $x \in C_{n+1}$.

Therefore , we may assume that A_{n+1} is a factor . With

this assumption, consider again $x \in A_{n+1} \cap C'_{n+1}$. If q is a minimal central projection of A_n, then $q \in C_n \subset C_{n+1}$, qA_n is a factor, $qA_{n+1}q$ is also a factor and, since q is central both in A_n and in $A'_n \cap A_{n+1}$, we have:

a') qC_n is a m.a.s.a. in qA_n.

b') qD_{n+1} is a m.a.s.a. in $(qA_n)' \cap (qA_{n+1}q)$.

c') $qx \in qA_{n+1}q$ commutes with $qC_{n+1} = \langle qC_n, qD_{n+1} \rangle$.

If from a'), b'), c') we infer that $qx \in qC_{n+1}$, then we have proved that $qx \in C_{n+1}$ for any minimal central projection q of A_n. Since 1 is a finite sum of minimal central projections of A_n, it follows that $x \in C_{n+1}$.

Therefore, in proving the inductive step, we may assume that A_{n+1} and A_n are both factors. But then it is clear that

$$A_{n+1} \simeq A_n \otimes (A'_n \cap A_{n+1}) \quad , \quad C_{n+1} \simeq C_n \otimes D_{n+1}$$

and, since C_n (resp. D_{n+1}) is a m.a.s.a. in A_n (resp. in $A'_n \cap A_{n+1}$), it follows obviously that C_{n+1} is a m.a.s.a. in A_{n+1}.

(iii) The equality we have to prove is obvious for $k = o$. Assuming that it is true for a fixed k, we get

$$\begin{aligned} C_{n+k+1} &= \langle C_{n+k}, D_{n+k+1} \rangle \\ &= \langle C_n, A'_n \cap C_{n+k}, D_{n+k+1} \rangle \\ &\subset \langle C_n, A'_n \cap C_{n+k+1}, A'_{n+k} \cap A_{n+k+1} \cap C_{n+k+1} \rangle \\ &\subset \langle C_n, A'_n \cap C_{n+k+1} \rangle \\ &\subset C_{n+k+1} \quad , \end{aligned}$$

which proves the desired equality by induction on k.

(ii) Let E be an abelian subalgebra such that

$$A_n' \cap C_{n+k} \subset E \subset A_n' \cap A_{n+k} .$$

Then

$$C_{n+k} = \langle C_n , A_n' \cap C_{n+k} \rangle \subset \langle C_n , E \rangle \subset A_{n+k} .$$

Since C_{n+k} is a m.a.s.a. in A_{n+k}, it follows that $E \subset C_{n+k}$ hence $E = A_n' \cap C_{n+k}$. Thus $A_n' \cap C_{n+k}$ is indeed a m.a.s.a. in $A_n' \cap A_{n+k}$.

Q.E.D.

I.1.2. Denote by $\{q_i\}_{i \in I_n}$ the minimal projections of C_n. For each $x \in A$ we define

$$P_n(x) = \sum_{i \in I_n} q_i \, x \, q_i .$$

It is clear that $P_n(x) \in C_n'$ and that the map $P_n : x \longmapsto P_n(x)$ is a projection of norm one of A onto C_n'. Thus, P_n is a conditional expectation of A with respect to C_n'.

Since C_n is a m.a.s.a. in A_n, we have $P(A_n) = C_n$. In fact,

$$P_n | A_n : A_n \longrightarrow C_n$$

<u>is the unique conditional expectation of</u> A_n <u>with respect to</u> C_n <u>and it is faithful</u>, i.e.

$$x \in A_n , \quad P_n(x^* x) = 0 \Longrightarrow x = 0 .$$

Of course, this is a well known fact. However, for the sake of completness and in order to establish some notations, we shall prove it. We may suppose that A_n is a factor. Then $\{q_i\}_{i \in I_n}$ is a complete set of mutually orthogonal and equivalent minimal projections of A_n , thus , for a fixed index $i_0 \in I_n$, we can find partial isometries $v_i \in A$ such that

(1) $v_i^* v_i = q_{i_0}$, $v_i v_i^* = q_i$, $v_i = v_i q_{i_0} = q_i v_i$; $i \in I_n$;

then an arbitrary element $x \in A_n$ is of the form

(2) $$x = \sum_{i,j \in I_n} \alpha_{ij} v_i v_j^* \quad , \quad \alpha_{ij} \in \mathbb{C} \quad .$$

If $\Phi : A_n \longrightarrow C_n$ is a conditional expectation , then

$$\Phi(v_i v_j^*) = \Phi(q_i(v_i v_j^*)q_j) = q_i q_j \Phi(v_i v_j^*) = \delta_{ij} q_i \quad ,$$

thus

$$\Phi(x) = \sum_{i,j} \alpha_{ij} \Phi(v_i v_j^*) = \sum_i \alpha_{ii} q_i = \sum_i q_i x q_i = P_n(x) \ .$$

Moreover ,

$$x^* x = \sum_{i,j} (\sum_k \overline{\alpha_{ki}} \alpha_{kj}) v_i v_j^* \quad , \quad P_n(x^* x) = \sum_h (\sum_k |\alpha_{kh}|^2) q_h$$

and therefore

$$P_n(x^* x) = 0 \Longrightarrow \alpha_{kh} = 0 \ , \ (\forall)\ k,h \Longrightarrow x = 0 \quad .$$

I.1.3. Now denote by $\{p_j\}_{j \in J_n}$ the minimal projections of D_{n+1} . Since $C_{n+1} = \langle C_n , D_{n+1} \rangle$, it follows that the minimal projections of C_{n+1} are the non-zero $q_i p_j$, $i \in I_n$, $j \in J_n$. Since $p_j \in D_{n+1} \subset A_n'$, for each $x \in A_n$ we have

$$\begin{aligned} P_{n+1}(x) &= \sum_{i \in I_n, j \in J_n} q_i p_j x q_i p_j = \\ &= \sum_{i \in I_n} q_i x q_i \sum_{j \in J_n} p_j = \\ &= \sum_{i \in I_n} q_i x q_i = P_n(x) \ . \end{aligned}$$

Therefore

$$P_{n+1} | A_n = P_n | A_n$$

and so

$$P_{n+k} \mid A_n = P_n \mid A_n \quad \text{for all } n \geqslant 0 , k \geqslant 0 .$$

Consequently , for $x \in \bigcup_{n=0}^{\infty} A_n$ we may define

$$P(x) = P_n(x) \quad \text{if} \quad x \in A_n .$$

Then

$$P : \bigcup_{n=0}^{\infty} A_n \ni x \longmapsto P(x) \in \bigcup_{n=0}^{\infty} C_n$$

is a projection of norm one .

We define

$$C = \langle \bigcup_{n=0}^{\infty} C_n \rangle \ (= \overline{\bigcup_{n=0}^{\infty} C_n})$$

and we denote again by

$$P : A \longrightarrow C$$

the unique bounded linear extension of $P : \bigcup_{n=0}^{\infty} A_n \longrightarrow \bigcup_{n=0}^{\infty} C_n$.

PROPOSITION . (i) C _is a m.a.s.a. in_ A , $A_n' \cap C$ _is a m.a.s.a. in_ A_n' _and , for each_ $n \geqslant 0$,

$$C \cap A_n = C_n \quad , \quad C = \langle C_n , A_n' \cap C \rangle .$$

(ii) $P : A \longrightarrow C$ _is a conditional expectation of_ A _with respect to_ C _and , for each_ $n \geqslant 0$,

$$P \circ P_n = P_n \circ P = P .$$

Proof . (i) Consider $x \in C'$. There is a sequence $x_n \in A_n$ such that $\lim_{n \to \infty} \|x_n - x\| = 0$. For each $n \geqslant 0$ we have $x \in C_n'$ and therefore $P_n(x) = x$. Thus ,

$$\|P_n(x_n) - x\| = \|P_n(x_n - x)\| \leqslant \|x_n - x\| .$$

It follows that $\lim_{n \to \infty} \|P_n(x_n) - x\| = 0$ and , since $P_n(x_n) \in C_n$, we get $x \in C$. Hence C is a m.a.s.a. in A .

Since $C \cap A_n$ is an abelian subalgebra of A_n which contains C_n and since C_n is a m.a.s.a. in A_n , it is clear that $C \cap A_n = C_n$.

By Lemma I.1.1.(iii) we have

$$C_{n+k} = \langle C_n , A_n' \cap C_{n+k} \rangle \subset \langle C_n , A_n' \cap C \rangle$$

for every $k \geq 0$. This obviously implies that $C = \langle C_n , A_n' \cap C \rangle$.

The fact that $A_n' \cap C$ is a m.a.s.a. in A_n' follows now as in the proof of I.1.1.(ii) .

(ii) Since $P_{n+k}(x) = P_n(x) = x$ for $x \in C_n$, we have $P(x) = x$ for $x \in \bigcup_{n=0}^{\infty} C_n$ and , by the continuity of P , we infer $P(x) = x$ for all $x \in C$. Thus , $P : A \longrightarrow C$ is a projection of norm one of A onto C , that is , a conditional expectation of A with respect to C .

Finally , for any $x \in A$ we have

$$P(P_n(x)) = P(\sum_{i \in I_n} q_i x q_i) = \sum_{i \in I_n} q_i P(x) = P(x) ,$$

$$P_n(P(x)) = \sum_{i \in I_n} q_i P(x) q_i = \sum_{i \in I_n} q_i P(x) = P(x) ,$$

where $\{q_i\}_{i \in I_n}$ are the minimal projections of C_n .

Q.E.D.

I.1.4. We now consider the set $\mathcal{U}_n$ consisting of all unitary elements $u \in A_n$ with the property

$$u^* C_n u = C_n .$$

Clearly , $\mathcal{U}_n$ is a group . If $u \in \mathcal{U}_n$, then $u^* C_n u = C_n$ and $u^* c u = c$ for any $c \in A_n' \cap C$. Since $C = \langle C_n , A_n' \cap C \rangle$, we get

(3) $$u^* C u = C .$$

Then, from $C_{n+1} = C \cap A_{n+1}$ we infer $u^* C_{n+1} u = C_{n+1}$. Thus,

$$\mathcal{U}_n \subset \mathcal{U}_{n+1} .$$

We define

$$\mathcal{U} = \bigcup_{n=0}^{\infty} \mathcal{U}_n .$$

Then $\mathcal{U}$ is a group of unitary elements of A and any $u \in \mathcal{U}$ satisfies the relation (3).

PROPOSITION. (i) $A = \text{c.l.m.}(\mathcal{U} C) = \text{c.l.m.}(C \mathcal{U})$.

(ii) $P(u^* x u) = u^* P(x) u$ <u>for all</u> $x \in A$, $u \in \mathcal{U}$.

<u>Proof</u>. (i) It suffices to show that

$$A_n = \text{l.m.}(\{uc ; u \in \mathcal{U}_n , c \in C_n\}) .$$

In order to avoid notational complications, we assume that A_n is a factor. Let $\{q_i\}_{i \in I_n}$ be the minimal projections of C_n and, as in Section I.1.2., denote by $\{v_i\}_{i \in I_n}$ the partial isometries satisfying the relations (1). Owing to relation (2) we see that it suffices to show that

$$v_i v_j^* \in \mathcal{U}_n C_n \quad , \quad i,j \in I_n .$$

But this is clear, since

$$v_i v_j^* = u_{ij} q_j$$

where

$$u_{ij} = 1 - (v_i - v_j)(v_i^* - v_j^*) \in \mathcal{U}_n \quad , \quad q_j \in C_n .$$

(ii) It is enough to prove the claim only for $x \in \bigcup_{n=0}^{\infty} A_n$. Since $\mathcal{U} = \bigcup_{n=0}^{\infty} \mathcal{U}_n$, we may suppose that there exists $n \geq 0$

such that $x \in A_n$ and $u \in \mathcal{U}_n$. Define

$$P'(y) = uP(u^*yu)u^* \quad \text{for} \quad y \in A_n \ .$$

Since $P(u^*yu) \in C_n$ and $u^*C_nu = C_n$, we have $P'(y) \in C_n$. It is easy to check that $P' : A_n \longrightarrow C_n$ is a conditional expectation of A_n with respect to C_n . By Section I.1.2. it follows that $P' = P_n \mid A_n = P \mid A_n$. Thus

$$uP(u^*xu)u^* = P(x) \ .$$

Q.E.D.

I.1.5. COROLLARY . If $x \in A_n'$, then $P(x) \in A_n'$.

Proof . For any $u \in \mathcal{U}_n$ we have $u^*xu = x$, thus , by I.1.4.(ii) , $u^*P(x)u = P(u^*xu) = P(x)$. Hence $P(x)$ commutes with any $u \in \mathcal{U}_n$. But $P(x)$ obviously commutes with any $c \in C_n$ and , since $A_n = \text{l.m.}(\mathcal{U}_nC_n)$, it follows that $P(x) \in A_n'$.

Q.E.D.

I.1.6. LEMMA . For any $n \geq 0$ we have :

$$\text{(i)} \qquad A_n' = \left\langle \bigcup_{k=0}^{\infty} A_n' \cap A_{n+k} \right\rangle ,$$

$$\text{(ii)} \quad A_n' \cap C = \left\langle \bigcup_{k=0}^{\infty} A_n' \cap C_{n+k} \right\rangle .$$

Proof . (i) Again , we shall assume that A_n is a factor and use the notations introduced in Section I.1.2. We define

$$Q_n(x) = \sum_{i \in I_n} v_i x v_i^* \quad \text{for all} \quad x \in A \ .$$

Then $Q_n(x)$ commutes with all $v_iv_j^*$, hence $Q_n(x) \in A_n'$. In fact , $Q_n : A \longrightarrow A_n'$ is a conditional expectation .

Consider $y \in A_n'$. There is a sequence $y_k \in A_{n+k}$ such that $\lim_{k\to\infty} \|y_k - y\| = 0$. Since $y \in A_n'$, we have $Q_n(y) = y$. Thus

$$\|Q_n(y_k) - y\| = \|Q_n(y_k - y)\| \leq \|y_k - y\|$$

and $\lim_{k\to\infty} \|Q_n(y_k) - y\| = 0$. But $Q_n(y_k) \in A_n' \cap A_{n+k}$, hence

$$y \in \langle \bigcup_{k=0}^{\infty} A_n' \cap A_{n+k} \rangle \quad .$$

(ii) By Corollary I.1.5. we have $P(A_n' \cap A_{n+k}) = A_n' \cap C_{n+k}$ and using (i) we obtain

$$P(A_n') = \langle \bigcup_{k=0}^{\infty} A_n' \cap C_{n+k} \rangle \quad .$$

Therefore , for every $c \in A_n' \cap C$,

$$c = P(c) \in \langle \bigcup_{k=0}^{\infty} A_n' \cap C_{n+k} \rangle \quad .$$

Q.E.D.

I.1.7. PROPOSITION . _If_ $x \in A_n$ _and_ $y \in A_n'$, _then_

$$P(xy) = P(x)P(y) \quad .$$

Proof . By Lemma I.1.6. it is sufficient to prove the equality of the statement only for $x \in A_n$ and $y \in A_n' \cap A_{n+k}$.

Thus , fix $n \geqslant 0$, $k \geqslant 0$, denote by $\{q_i\}_{i \in I_n}$ the minimal projections of C_n and by $\{p_j\}_{j \in J_{n,k}}$ the minimal projections of $A_n' \cap C_{n+k}$. By Lemma I.1.1.(iii) it follows that the non-zero $q_i p_j$, $i \in I_n$, $j \in J_{n,k}$, are the minimal projections of C_{n+k} . We define

$$P_{n+k/n}(z) = \sum_{j \in J_{n,k}} p_j z p_j \in (A_n' \cap C_{n+k})' \quad \text{for all} \quad z \in A \; .$$

Then $P_{n+k/n} : A \longrightarrow (A_n' \cap C_{n+k})'$ is a conditional expectation, in particular

$$P_{n+k/n}(xy) = x\, P_{n+k/n}(y) \quad \text{for all} \quad x \in A_n \text{ , } y \in A \text{ .}$$

As in Sections I.1.2.,I.1.3. we see that $P_{n+k/n}$ is the unique conditional expectation of $A_n' \cap A_{n+k}$ with respect to $A_n' \cap C_{n+k}$ and

$$P_{n+k/n} \big| A_n' \cap A_{n+k} \;=\; P_{n+k} \big| A_n' \cap A_{n+k} \quad .$$

Moreover , for any $z \in A$,

$$P_n(P_{n+k/n}(z)) = P_n(\sum_j p_j z p_j) = \sum_{i,j} q_i p_j x p_j q_i = P_{n+k}(z) \quad ,$$

therefore

$$P_n \circ P_{n+k/n} = P_{n+k/n} \circ P_n = P_{n+k} \quad .$$

Thus , for $x \in A_n$ and $y \in A_n' \cap A_{n+k}$, we have :

$$\begin{aligned} P(xy) &= P_{n+k}(xy) \\ &= P_n(P_{n+k/n}(xy)) \\ &= P_n(x\, P_{n+k/n}(y)) \\ &= P_n(x)P_{n+k/n}(y) \\ &= P_{n+k}(x)P_{n+k}(y) \quad = \quad P(x)P(y) \end{aligned}$$

Q.E.D.

It can be proved that

$$A_n' \cap C_{n+k+1} \quad = \quad \langle A_n' \cap C_{n+k} \text{ , } A_{n+k}' \cap C_{n+k+1} \rangle$$

which implies

$$P_{n+k/n} \big| A_{n+k} \;=\; P_{n+k+1/n} \big| A_{n+k} \quad .$$

Thus , we may define a map $P_{\infty/n} : A \longrightarrow (A_n' \cap C)'$ by

$$P_{\infty/n}(x) = P_{n+k/n}(x) \quad \text{if} \quad x \in A_{n+k} \text{ .}$$

This map is a conditional expectation , $P_{\infty/n} \big| A_n' \;=\; P \big| A_n'$ and

$$P \;=\; P_{\infty/n} \circ P_n \;=\; P_n \circ P_{\infty/n} \quad .$$

I.1.8. In this section we shall determine suitable systems of matrix units for the finite-dimensional C^*- algebras A_n .

Consider

$$A_n = \sum_{k \in K_n} A_n^k$$

the decomposition of A_n in factor components and denote by $\left\{q_i^{(n)}\right\}_{i \in I_n^k}$ the minimal projections of $A_n^k \cap C_n$. For each $k \in K_n$ there is a system of matrix units for the factor A_n^k with respect to the m.a.s.a. $A_n^k \cap C_n$, that is a set

$$\left\{ e_{ij}^{(n)} \; ; \; i,j \in I_n^k \right\}$$

consisting of partial isometries

$$e_{ij}^{(n)} : q_j^{(n)} \sim q_i^{(n)}$$

such that

$$e_{ij}^{(n)} e_{rs}^{(n)} = \delta_{jr} e_{is}^{(n)} \quad , \quad e_{ij}^{(n)*} = e_{ji}^{(n)} \quad .$$

Such a system is completely determined once we choose an index $i_o \in I_n^k$ and the partial isometries $v_i = e_{ii_o}^{(n)}$, $i \in I_n^k$, since

$$e_{ij}^{(n)} = v_i v_j^* \quad , \quad i,j \in I_n^k \quad .$$

The whole system of matrix units

$$\left\{ e_{ij}^{(n)} \; ; \; i,j \in I_n^k \, , \, k \in K_n \right\}$$

is a linear basis for A_n . If $i,j \in I_n^k$, $r,s \in I_n^h$ and $h \neq k$, then $e_{ij}^{(n)} e_{rs}^{(n)} = 0$.

PROPOSITION . <u>The systems</u> $\left\{ e_{ij}^{(n)} \; ; \; i,j \in I_n^k \, , \, k \in K_n \right\}$ <u>of matrix units for</u> A_n <u>with respect to</u> C_n <u>can be chosen such that , for every</u> $n \geq o$, <u>the following assertion holds :</u>

(4) <u>each</u> $e_{ij}^{(n)}$ <u>is a sum of some</u> $e_{rs}^{(n+1)}$.

<u>Proof</u> . We proceed by induction . Let $\{e^{(n)}_{i_1 i_2}\}$ be the system of matrix units for A_n with respect to C_n and let $\{f_{j_1 j_2}\}$ be some system of matrix units of $A'_n \cap A_{n+1}$ with respect to D_{n+1} . The non-zero $e^{(n)}_{ii} f_{jj}$ are the minimal projections of C_{n+1} and the non-zero $e^{(n)}_{i_1 i_2} f_{j_1 j_2}$ are partial isometries between such projections . Moreover , for every i_1, i_2 ,

$$e^{(n)}_{i_1 i_2} = \sum_j e^{(n)}_{i_1 i_2} f_{jj} \quad .$$

Now we may take as $\{e^{(n+1)}_{rs}\}$ any system of matrix units of A_{n+1} containing the $e^{(n)}_{i_1 i_2} f_{j_1 j_2}$'s .

Q.E.D.

I.1.9. There is a homomorphism of the group $\mathcal{U}$ onto a group Γ of $*$ - automorphisms of C , namely , for $u \in \mathcal{U}$, the corresponding $*$ - automorphism $\gamma_u \in \Gamma$ is

$$\gamma_u : C \ni c \longmapsto u^* c u \in C \quad .$$

The kernel of this homomorphism is easily seen to be

$$\mathcal{U} \cap C = \bigcup_{n=0}^{\infty} \mathcal{U}_n \cap C_n \quad .$$

For given systems of matrix units $\{e^{(n)}_{ij} ;\ i,j \in I^k_n \ ,\ k \in K_n\}$ satisfying condition (4) of I.1.8., we shall construct a subgroup U of $\mathcal{U}$ such that $\mathcal{U}$ be the semi-direct product of its normal subgroup $\mathcal{U} \cap C$ by U .

Let U_n be the subgroup of $\mathcal{U}_n$ consisting of all

$$u = \sum_{k \in K_n} \sum_{i \in I^k_n} e^{(n)}_{i,\sigma_k(i)}$$

where , for each $k \in K_n$, σ_k is some permutation of I^k_n . Thus, U_n is generated by the

$$u_{ij}^{(n)} = 1 - e_{ii}^{(n)} - e_{jj}^{(n)} + e_{ij}^{(n)} + e_{ji}^{(n)}$$

with $i,j \in I_n^k$, $k \in K_n$. Remark that

$$e_{ij}^{(n)} = u_{ij}^{(n)} q_j^{(n)} = q_i^{(n)} u_{ij}^{(n)} .$$

It is easily verified that U_n is the set of all $u \in \mathcal{U}_n$ of the form

$$u = \sum_{k \in K_n} \sum_{i,j \in I_n^k} \alpha_{ij} e_{ij}^{(n)} , \quad \alpha_{ij} \in \{0,1\} \text{ for all } i,j ,$$

and that $\mathcal{U}_n$ is the semi-direct product of $\mathcal{U}_n \cap C_n$ by U_n .

Thus , $U_n \subset U_{n+1}$ and putting

$$U = \bigcup_{n=0}^{\infty} U_n ,$$

it follows that U is a subgroup of $\mathcal{U}$ and $\mathcal{U}$ is the semi-direct product of $\mathcal{U} \cap C$ by U . Moreover , U and Γ are isomorphic and since

$$A_n = \text{l.m.}(U_n C_n) = \text{l.m.}(C_n U_n) ,$$

we have

$$A = \text{c.l.m.}(UC) = \text{c.l.m.}(CU) .$$

I.1.10. We now denote by Ω the Gelfand spectrum of the commutative C^*- algebra C . Then Ω is a compact topological space , $C \simeq C(\Omega)$ and we may view Γ as a group of homeomorphisms of Ω . Thus , we obtain a topological dynamical system

$$(\Omega, \Gamma)$$

associated to the given AF - algebra A .

Consider the Hilbert space $\ell^2(\Omega)$ with orthonormal basis $\{t \; ; \; t \in \Omega\}$ and denote by $(\cdot \mid \cdot)$ the scalar product .

Each function $f \in C(\Omega)$ defines a "multiplication operator" T_f on $\ell^2(\Omega)$ by

$$T_f(h) = fh \qquad ; \quad h \in \ell^2(\Omega) .$$

On the other hand, each element $\gamma \in \Gamma$ defines a "permutation operator" V_γ on $\ell^2(\Omega)$ by

$$V_\gamma(h)(t) = h(\gamma^{-1}(t)) , \quad t \in \Omega \quad ; \quad h \in \ell^2(\Omega) .$$

Let us denote by

$$A(\Omega, \Gamma)$$

the C^*- algebra generated in $L(\ell^2(\Omega))$ by the operators T_f, $f \in C(\Omega)$, and V_γ, $\gamma \in \Gamma$.

THEOREM. <u>Given an arbitrary</u> AF - <u>algebra</u> A <u>there exist</u>

a) <u>a m.a.s.a.</u> C <u>in</u> A,

b) <u>a conditional expectation</u> P <u>of</u> A <u>with respect to</u> C,

c) <u>a subgroup</u> U <u>of the unitary group of</u> A,

<u>such that</u>

(i) $u^* C u = C$ <u>for all</u> $u \in U$,

(ii) $P(u^*xu) = u^*P(x)u$ <u>for all</u> $u \in U$, $x \in A$,

(iii) $A = \text{c.l.m.}(UC) = \text{c.l.m.}(CU)$.

<u>Moreover, let</u> Ω <u>be the Gelfand spectrum of</u> C <u>and</u> Γ <u>be the group of homeomorphisms of</u> Ω <u>induced by</u> U. <u>Then there is a</u> $*$ - <u>isomorphism</u>

$$A \simeq A(\Omega, \Gamma)$$

<u>such that</u>

$$P(x)(t) = (xt \mid t) , \quad t \in \Omega ; \quad x \in A .$$

<u>Proof</u>. The first part of the Theorem was already proved

in the preceding Sections . It remains to construct a $*$ - isomorphism between A and $A(\Omega,\Gamma)$ with the stated property .

Fix $n \geqslant 0$, denote by $\gamma_{ij}^{(n)}$ the image of $u_{ij}^{(n)} \in U_n \subset U$ under the map $U \longrightarrow \Gamma$ and by $f_j^{(n)}$ the function on Ω corresponding to $q_j^{(n)} \in C_n \subset C$ and put

$$T_j^{(n)} = T_{f_j^{(n)}} \quad , \quad V_{ij}^{(n)} = V_{\gamma_{ij}^{(n)}} \quad ; \quad i,j \in I_n^k \, , \; k \in K_n \; .$$

Then the correspondence

$$e_{ij}^{(n)} = u_{ij}^{(n)} q_j^{(n)} \longmapsto V_{ij}^{(n)} T_j^{(n)} \quad ; \quad i,j \in I_n^k \, , \; k \in K_n \, ,$$

has a unique linear extension up to an isometric $*$ - homomorphism

$$\rho^{(n)} : A_n \longrightarrow A(\Omega,\Gamma)$$

and , identifying an element of A_n with its image under $\rho^{(n)}$ and a point of Ω with the corresponding vector in $\ell^2(\Omega)$, we have

$$P_n(x)(t) = (xt \mid t) \quad , \quad t \in \Omega \quad ; \quad x \in A_n \; .$$

Since the systems of matrix units were chosen with property (4) , it follows that

$$\rho^{(n+1)} \mid A_n = \rho^{(n)} \quad .$$

We thus obtain an isometric $*$ - homomorphism of $\bigcup_{n=0}^{\infty} A_n$ into $A(\Omega,\Gamma)$ which clearly extends to a $*$ - isomorphism

$$A \simeq A(\Omega,\Gamma)$$

having the desired property .

Q.E.D.

Note that , although the concrete C^*- algebra $A(\Omega,\Gamma)$ and its conditional expectation with respect to $C(\Omega)$ depend only on the choice of the minimal projections , the constructed $*$ - isomorphism between A and $A(\Omega,\Gamma)$ is essentially based on a suitable choice of the complete systems of matrix units .

I.1.11. For later use , we shall give a convenient description of Ω as well as of the action of Γ on Ω .

The Gelfand spectrum Ω_n of the commutative C^*- algebra C_n can be identified with the finite set $\{q_i^{(n)}\}_{i\in I_n}$ of all the minimal projections of C_n . The map $\Omega_{n+1} \longrightarrow \Omega_n$ corresponding to the inclusion map $C_n \longrightarrow C_{n+1}$ associates to every minimal projection of C_{n+1} the unique minimal projection of C_n containing it . Since the C^*- algebra C is the direct limit of the C^*- algebras C_n following the inclusion maps , it follows that Ω can be identified with the topological inverse limit of the discrete spaces Ω_n following the maps $\Omega_{n+1} \longrightarrow \Omega_n$.

Therefore , the points of Ω can be represented as sequences

$$t = (q_t^{(o)} \geqslant q_t^{(1)} \geqslant \dots \geqslant q_t^{(n)} \geqslant q_t^{(n+1)} \geqslant \dots)$$

where $q_t^{(n)}$ is a minimal projection of C_n , for all $n \geqslant o$.

A point $t \in \Omega$ is adherent to a set $\omega \subset \Omega$ if and only if , for each $n \geqslant o$, there exists $s \in \omega$ such that

$$q_t^{(n)} = q_s^{(n)} \quad ;$$

remark that the last equality implies

$$q_t^{(k)} = q_s^{(k)} \qquad \text{for all} \quad o \leqslant k \leqslant n \ .$$

Finally , for any $u \in \mathcal{U}$ and any $t \in \Omega$, the point $\gamma_u(t) \in \Omega$ is determined by

$$q^{(n)}_{\gamma_u(t)} = u^* q^{(n)}_t u \quad , \qquad n \geqslant 0 \quad .$$

I.1.12. O.Bratteli ([1]) has studied AF - algebras by means of a certain partially ordered set called the diagram of the AF - algebra , which reflects the factor components of the A_n ' s and their partial embeddings . These diagrams are particularly useful for constructing approximately finite dimensional C^*- algebras . For instance , O.Bratteli ([4]) proved that any separable commutative C^*- algebra is the center of some AF - algebra .

The diagonalisation of AF - algebras presented here is closely patterned after a similar method first used in the study of the canonical anticommutation relations of mathematical physics by L.Gårding and A.Wightman ([12]) .

§ 2 Ideals in AF - algebras

In this section we study the closed two sided ideals of an AF - algebra A and we interpret the results in terms of the topological dynamical system associated to a diagonalisation of A .

I.2.1. We begin with a known result ([1], 3.1.) :

LEMMA . Let A be a C^*- algebra and A_n be an ascending sequence of C^*- subalgebras such that

$$A = \langle \bigcup_{n=0}^{\infty} A_n \rangle \quad .$$

Then for any closed two sided ideal J *of* A *we have*

$$J = \langle \bigcup_{n=0}^{\infty} J \cap A_n \rangle .$$

Proof . The canonical $*$ - homomorphisms $A_n/J \cap A_n \longrightarrow A/J$ are injective and therefore isometric . For any $x \in J$ there is a sequence $x_n \in A_n$ with $\lim_{n\to\infty} \|x_n - x\| = 0$. It follows that $\lim_{n\to\infty} \|x_n/J\| = 0$, hence $\lim_{n\to\infty} \|x_n/J \cap A_n\| = 0$. Thus there is a sequence $y_n \in J \cap A_n$ such that $\lim_{n\to\infty} \|x_n - y_n\| = 0$ and this entails $\lim_{n\to\infty} \|x - y_n\| = 0$.

Q.E.D.

I.2.2. Now let $A = \langle \bigcup_{n=0}^{\infty} A_n \rangle$ be an AF - algebra with m.a.s.a. $C = \langle \bigcup_{n=0}^{\infty} C_n \rangle$, conditional expectation $P : A \longrightarrow C$ and group $\mathcal{U}$ constructed as in § 1 . Denote by Γ the corresponding group of $*$ - automorphisms of C .

If J is any closed two sided ideal of A , then

$$I_J = J \cap C$$

is a Γ - stable closed ideal of C . Conversely , using standard arguments ([6]) , we shall prove the following

LEMMA . *Let* I *be a* Γ - *stable closed ideal of* C . *Then*

$$J(I) = \{x \in A \; ; \; P(x^*x) \in I\}$$

is a closed two sided ideal of A .

Proof . Clearly , $J(I)$ is a closed subset of A .

Since $(x + y)^*(x + y) \leqslant 2(x^*x + y^*y)$, from $x , y \in J(I)$ we infer

$$P((x + y)^*(x + y)) \leqslant 2(P(x^*x) + P(y^*y)) \in I ,$$

thus $P((x + y)^*(x + y)) \in I$, i.e. $x + y \in J(I)$.

Since $(ax)^*(ax) \leqslant \|a\|^2 x^*x$, from $x \in J(I)$ and $a \in A$ we infer

$$P((ax)^*(ax)) \leqslant \|a\|^2 P(x^*x) \in I ,$$

thus $P((ax)^*(ax)) \in I$, i.e. $ax \in J(I)$.

In order to show that $x \in J(I)$, $a \in A \Longrightarrow xa \in J(I)$, it suffices to consider $a = uc$ with $u \in \mathcal{U}$, $c \in C$ (see I.1.4.(i)). Then , by I.1.4.(ii) , we have

$$P((xa)^*(xa)) = P(c^*u^*x^*xuc) = c^*c\ u^*P(x^*x)u .$$

Since $P(x^*x) \in I$ and I is Γ- stable , it follows that $u^*P(x^*x)u \in I$ and therefore $P((xa)^*(xa)) \in I$, i.e. $xa \in J(I)$.

Q.E.D.

I.2.3. Let J be a closed two sided ideal of the AF - algebra A and $x \in A$. Then

LEMMA . $x \in J \Longrightarrow P(x) \in J$.

<u>Proof</u> . Clearly , for each $n \geqslant 0$ we have

$$x \in J \Longrightarrow P_n(x) \in J .$$

Since $P(x) = \lim_{n \to \infty} P_n(x)$, the Lemma follows .

Q.E.D.

I.2.4. THEOREM . <u>For any closed two sided ideal</u> J <u>of the</u> AF - <u>algebra</u> A <u>we have</u>

$$J = J(I_J) = \left\{x \in A \ ; \ P(x^*x) \in J \cap C\right\} .$$

<u>Thus</u> , $J \longmapsto I_J$ <u>and</u> $I \longmapsto J(I)$ <u>are inverse to one another correspondences between the set of all closed two sided ideals</u> J

<u>of</u> A <u>and the set of all</u> Γ - <u>stable closed ideals</u> I <u>of</u> C .

<u>Proof</u> . If $x \in J$, then $x^*x \in J$ hence , by I.2.3. , $P(x^*x) \in J \cap C = I_J$, i.e. $x \in J(I_J)$.

Conversely , by I.2.1. and I.2.2. , we have

$$J = \left\langle \bigcup_{n=0}^{\infty} J \cap A_n \right\rangle ,$$

$$J(I_J) = \left\langle \bigcup_{n=0}^{\infty} J(I_J) \cap A_n \right\rangle =$$

$$= \left\langle \bigcup_{n=0}^{\infty} \{x \in A_n \; ; \; P_n(x^*x) \in (J \cap A_n) \cap C_n\} \right\rangle$$

and therefore it suffices to show that

$$J \cap A_n = \{x \in A_n \; ; \; P_n(x^*x) \in (J \cap A_n) \cap C_n\} .$$

Since $J \cap A_n$ is a two sided ideal of A_n , there is a central projection p of A_n such that $J \cap A_n = pA_n$. Consider

$$x \in A_n \quad \text{with} \quad P_n(x^*x) \in J .$$

Then $x = px + (1 - p)x$ and

$$pA_n \ni P_n(x^*x) = P_n(px^*x + (1 - p)x^*x) = pP_n(x^*x) + (1 - p)P_n(x^*x)$$

whence

$$P_n(((1 - p)x)^*((1 - p)x)) = 0 .$$

Using I.1.2. we infer that

$$(1 - p)x = 0$$

and therefore

$$x = px \in pA_n = J \cap A_n .$$

The equality $I = I_{J(I)}$ is obvious for any Γ - stable closed ideal of C .

Q.E.D.

Remark that the correspondences $J \longmapsto I_J$ and $I \longmapsto J(I)$ are increasing with respect to inclusion .

I.2.5. COROLLARY . Let J_1 and J_2 be closed two sided ideals of A . Then

$$J_1 = J_2 \Longleftrightarrow J_1 \cap C = J_2 \cap C \ .$$

Proof . Follows obviously from Theorem I.2.4.

Q.E.D.

I.2.6. COROLLARY . Let J be any closed two sided ideal of A . Then

J = the closed two sided ideal of A generated by $J \cap C$.

Proof . Denote by J_1 the closed two sided ideal of A generated by $J \cap C$. Clearly , $J_1 \subset J$. Since

$$J_1 \cap C \subset J \cap C \subset J_1 \cap C \ ,$$

the equality $J = J_1$ follows from I.2.5.

Q.E.D.

I.2.7. COROLLARY . The conditional expectation $P : A \longrightarrow C$ is faithful :

$$x \in A \ , \ P(x^*x) = 0 \implies x = 0 \ .$$

Proof . Follows from Theorem I.2.4. applied to $J = 0$.

Q.E.D.

I.2.8. THEOREM . For any primitive ideal J of the AF - algebra A there is a maximal ideal I of C such that

$$J \cap C = \bigcap_{u \in \mathcal{U}} u^* I u \quad .$$

The proof will be given in Section I.2.10.

Remark that Theorem I.2.4. further implies

$$J = \left\{ x \in A \; ; \; P(x^*x) \in u^* I u \; , \; (\forall) \; u \in \mathcal{U} \right\} .$$

I.2.9. Denote by Ω the Gelfand spectrum of C and consider Γ as a group of homeomorphisms of Ω .

There is a one-to-one correspondence between the closed ideals I of $C \simeq C(\Omega)$ and the closed subsets ω of Ω , which is given by

$$\omega_I = \left\{ t \in \Omega \; ; \; f(t) = 0 \; , \; (\forall) \; f \in I \right\} \quad ,$$

$$I_\omega = \left\{ f \in C \; ; \; f(t) = 0 \; , \; (\forall) \; t \in \Omega \right\} \quad .$$

This correspondence is decreasing with respect to inclusion .

The closed ideal I of C is Γ - stable if and only if the subset ω_I of Ω is Γ - stable . Owing to Theorem I.2.4. it follows that

$$J \longmapsto \omega_{J \cap C}$$

is a decreasing one-to-one correspondence between the closed two sided ideals of A and the Γ - stable closed subsets of Ω .

A closed two sided ideal J of A is called <u>primitive</u> if it is the kernel of an irreducible representation of A . It is known ([5]) that J is primitive if and only if , for any closed two sided ideals J_1 and J_2 of A , the following implication holds :

$$J = J_1 \cap J_2 \Longrightarrow \text{either } J = J_1 \text{ or } J = J_2 \, .$$

A Γ - stable closed subset ω of Ω will be called Γ - irreducible if , for any Γ - stable closed subsets ω_1 and ω_2 of Ω , the following implication holds :

$$\omega = \omega_1 \cup \omega_2 \Longrightarrow \text{either } \omega = \omega_1 \text{ or } \omega = \omega_2 \, .$$

Thus , the correspondence

$$J \longmapsto \omega_{J \cap C}$$

carries the primitive ideals of A onto the Γ - stable , Γ - irreducible closed subsets of Ω .

Let us denote by $\Gamma(t)$ the Γ - orbit of $t \in \Omega$ and by $\overline{\Gamma(t)}$ its closure . Then Theorem I.2.8. rephrases as follows

THEOREM . For any primitive ideal J of the AF - algebra A there is a point $t_o \in \Omega$ such that

$$\omega_{J \cap C} = \overline{\Gamma(t_o)} \, .$$

This entails the following property of the topological dynamical system (Ω, Γ) :

COROLLARY . The Γ - stable Γ - irreducible closed subsets of Ω coincide with the closures of the Γ - orbits .

The set $\omega_{J \cap C}$ associated to a closed two sided ideal J of A has a simple description in the terms explained in Section I.1.11. Namely , let π be a representation of A with kernel J. Then the set $\omega_{J \cap C}$ consists of all points $t \in \Omega$ having the property

$$\pi(q_t^{(n)}) \neq 0 \quad \text{for all} \quad n \geq 0 \; .$$

I.2.10. For the proof of Theorem I.2.8. we need two Lemmas.

Let π be a factor representation of A on the Hilbert space H such that

$$\ker \pi = J \; .$$

LEMMA 1 . Let e_1 , e_2 be projections of A_n such that

$$\pi(e_1) \neq 0 \quad , \quad \pi(e_2) \neq 0 \quad .$$

Then there exist $k \geq n$ and a minimal central projection p of A_k such that

$$\pi(pe_1) \neq 0 \quad , \quad \pi(pe_2) \neq 0 \quad .$$

Proof . Indeed , suppose the contrary holds . Then , for every $k \geq n$, there exist mutually orthogonal central projections $p_1^{(k)}$, $p_2^{(k)}$ of A_k with

(1) $$p_1^{(k)} + p_2^{(k)} = 1 \; ,$$

(2) $$\pi(p_1^{(k)}e_1) = 0 \quad , \quad \pi(p_2^{(k)}e_2) = 0 \; .$$

Since the unit ball of $L(H)$ is wo - compact , we may assume that the sequences $\{\pi(p_1^{(k)})\}$, $\{\pi(p_2^{(k)})\}$ are wo - convergent . Denote by p_1 , p_2 their corresponding limits . Then p_1 , p_2 are positive operators contained in the center of the von Neumann factor generated by $\pi(A)$ in $L(H)$, therefore they are scalar operators

$$p_1 = \lambda_1 \quad , \quad p_2 = \lambda_2 \quad ; \quad \lambda_1 , \lambda_2 \in [0,\infty).$$

Now from (1) we infer $\lambda_1 + \lambda_2 = 1$, while (2) implies that

$\lambda_1 = \lambda_2 = 0$. This contradiction proves the Lemma .

Q.E.D.

LEMMA 2 . <u>There is a sequence</u> $\{p^{(n)}\}$ <u>of minimal central projections</u> $p^{(n)}$ <u>of</u> A_n <u>with the properties</u>

(i) $\pi(p^{(1)} \dots p^{(n)}) \neq 0$ <u>for all</u> $n \geq 1$;

(ii) <u>for any minimal projection</u> q <u>of</u> C_n <u>with</u> $\pi(q) \neq 0$ <u>there exists</u> $k \geq n$ <u>such that</u> $\pi(p^{(k)}q) \neq 0$.

<u>Proof</u> . Indeed , let us write the set

$$\bigcup_{n=1}^{\infty} \{q \; ; \; q \text{ is a minimal projection of } C_n \text{ and } \pi(q) \neq 0\}$$

as a sequence $\{e_1 , e_2 , \dots , e_j , \dots\}$. Owing to Lemma 1 , we find by induction a sequence $\{p^{(k_j)}\}$ of minimal central projections $p^{(k_j)}$ of A_{k_j} such that

$$k_j < k_{j+1} ,$$

$$\pi(p^{(k_j)} p^{(k_{j-1})} \dots p^{(k_1)}) \neq 0 ,$$

$$\pi(p^{(k_j)} e_j) \neq 0 .$$

Clearly , this sequence can be refined up to a sequence $\{p^{(n)}\}$ having the stated properties .

Q.E.D.

<u>Proof of Theorem</u> I.2.8. Put $\omega = \omega_{J \cap C}$ and choose a sequence $\{p^{(n)}\}$ as in Lemma 2 . The condition (i) satisfied by the $p^{(n)}$'s and the compacity of Ω entail the existence of point $t_0 \in \Omega$ such that

$$p^{(n)}(t_0) \neq 0 \quad \text{for all} \quad n \geq 1 .$$

This means that

$$q^{(n)}_{t_o} \leqslant p^{(n)} \quad \text{for all} \quad n \geqslant 1 \quad ,$$

the notation being as in Section I.1.11. Therefore , $p^{(n)}$ is the central support of the minimal projection $q^{(n)}_{t_o}$ in A_n . Since $\pi(p^{(n)}) \neq 0$, it follows that

$$\pi(q^{(n)}_{t_o}) \neq 0 \quad \text{for all} \quad n \geqslant 1 \quad .$$

Thus , $t_0 \in \omega$ and consequently $\overline{\Gamma(t_0)} \subset \omega$.

Now consider $t \in \omega$ and fix $n \geqslant 1$. The condition (ii) satisfied by the $p^{(n)}$'s shows that there exists $k_n \geqslant n$ such that

$$p^{(k_n)} q^{(n)}_{t} \neq 0 \quad .$$

Therefore , there is a minimal projection $r^{(k_n)}$ of C_{k_n} with central support $p^{(k_n)}$ in A_{k_n} such that

$$r^{(k_n)} q^{(n)}_{t} \neq 0 \quad .$$

Since $q^{(k_n)}_{t_o}$ is also a minimal projection of C_{k_n} with central support $p^{(k_n)}$ in A_{k_n} , there exists $u \in \mathcal{U}_{k_n}$ such that

$$r^{(k_n)} = u^* q^{(k_n)}_{t_o} u \quad .$$

Thus ,

$$u^* q^{(k_n)}_{t_o} u \leqslant q^{(n)}_{t} \quad .$$

On the other hand ,

$$u^* q^{(k_n)}_{t_o} u \leqslant u^* q^{(n)}_{t_o} u \quad .$$

Since $q^{(n)}_{t}$ and $u^* q^{(n)}_{t_o} u$ are both minimal projections in A_n , it follows that

$$q_t^{(n)} = u^* q_{t_o}^{(n)} u \quad .$$

We have proved that, for each $n \geqslant 1$, there exists $s \in \Gamma(t_o)$ such that

$$q_t^{(n)} = q_s^{(n)} \quad .$$

This means that $t \in \overline{\Gamma(t_o)}$.

Therefore,

$$\omega = \overline{\Gamma(t_o)} \quad ,$$

which proves Theorem I.2.9. and its equivalent form, Theorem I.2.8.

Q.E.D.

The above proof shows that the kernel of any factor representation of the AF - algebra A is a primitive ideal, but this result is known for all separable C^*- algebras ([5]).

On the other hand, the same proof shows that any primitive ideal of the AF - algebra $A = \langle \bigcup_{n=o}^{\infty} A_n \rangle$ is the kernel of a direct limit representation of irreducible representations of the A_n's.

I.2.11. The primitive spectrum Prim(A) of A is the set of all primitive ideals of A endowed with the hull-kernel topology. The preceding results show that Prim(A) can be identified with the set of all closures of Γ- orbits. Defining an equivalence relation " $\sim$ " on Ω by

$$t_1 \sim t_2 \Longleftrightarrow \overline{\Gamma(t_1)} = \overline{\Gamma(t_2)} \quad ,$$

it can be easily verified that Prim(A) is homeomorphic with the quotient space $\Omega/\sim$ endowed with the quotient topology.

I.2.12. In his approach to AF - algebras based on diagrams, O.Bratteli has also studied the closed two sided ideals. Instead of considering the intersections of the ideals with the m.a.s.a. C, O.Bratteli considers the intersections with the smaller abelian subalgebra generated by the centers of the A_n's, the results being quite similar (see [1], 3.3., 3.8. and [2], 5.1.). His approach is particularly well adapted for problems such as the determination of all topological spaces which are spectra of AF - algebras (see [2], 4.2. and [3]).

§ 3 Some representations of AF - algebras

We consider an AF - algebra $A = \langle \bigcup_{n=0}^{\infty} A_n \rangle$ together with the m.a.s.a. C, the conditional expectation $P : A \longrightarrow C$ and the group U as in § 1. Let (Ω, Γ) be the associated topological dynamical system and $\mathcal{B}$ the sigma-algebra of Borel subsets of Ω. In this section we shall study two kinds of representations of A, π_μ and ρ_μ, associated with Γ- quasi-invariant measures μ on the Borel space $(\Omega, \mathcal{B})$.

A positive measure on Ω will always mean a positive regular Borel measure on Ω. A probability measure μ on Ω is a positive measure of mass 1, i.e. $\mu(\Omega) = 1$. Two positive measures μ, ν on Ω are equivalent if μ is absolutely continuous with respect to ν and ν is absolutely continuous with respect to μ, that is if μ and ν have the same null-sets.

For a positive measure μ on Ω and a homeomorphism γ of Ω onto Ω we shall denote by μ^{γ} the transform of μ by γ. Then μ is Γ- invariant (resp. Γ- quasi-invariant) if $\mu^{\gamma} = \mu$ (resp. μ^{γ} equivalent to μ) for all $\gamma \in \Gamma$. The positive measure μ is Γ- ergodic if the only Γ- invariant elements of $L^{\infty}(\Omega, \mu)$ are the scalars.

I.3.1. The construction of the representations π_{μ}.

Let μ be a Γ- quasi-invariant probability measure on Ω. Then μ can be regarded as a state of the commutative C^*- algebra $C \simeq C(\Omega)$ and therefore

$$\varphi_{\mu} = \mu \circ P$$

is a state of A. The Gelfand-Naimark-Segal (abreviated GNS) construction associates to φ_{μ} a representation π_{μ} of A on a Hilbert space H_{μ} and a cyclic unit vector $\xi_{\mu} \in H_{\mu}$ for π_{μ} such that

$$\varphi_{\mu}(x) = (\pi_{\mu}(x)\xi_{\mu} \mid \xi_{\mu}) \quad , \quad x \in A \quad .$$

For the von Neumann algebra generated by $\pi_{\mu}(A)$ in $L(H_{\mu})$ we use the bicommutant notation, $\pi_{\mu}(A)''$.

Since μ is Γ- quasi-invariant, its support Ω_{μ} is a

Γ- stable closed subset of Ω . Then it follows from § 2 that

$$J_\mu = \left\{x \in A \ ; \ P(x^*x)(t) = 0 \ , \ (\forall)\ t \in \Omega_\mu\right\}$$

is a closed two sided ideal of A .

Clearly , μ as a state of C is faithful if and only if $\Omega_\mu = \Omega$ and it is Γ- invariant if and only if the measure μ on Ω is Γ- invariant .

I.3.2. Let us recall that a state φ of A is central if

$$\varphi(xy) = \varphi(yx) \quad , \quad x , y \in A \quad .$$

PROPOSITION . _A state_ φ _of_ A _is central if and only if there exists a_ Γ - _invariant state_ μ _of_ C _such that_

$$\varphi = \mu \circ P \quad .$$

In this case $\mu = \varphi | C$.

Proof . If φ is central , then $\varphi | C$ is clearly Γ- invariant . Moreover , for fixed $n \geq 0$, denoting by $\{q_i\}_{i \in I_n}$ the minimal projections of C_n , we have

$$\varphi(P_n(x)) = \varphi(\sum_{i \in I_n} q_i x q_i) = \varphi(\sum_{i \in I_n} x q_i) = \varphi(x) \quad , \quad x \in A \ .$$

Hence

$$\varphi(x) = \lim_{n \to \infty} \varphi(P_n(x)) = \varphi(P(x)) \quad , \quad x \in A \quad .$$

Conversely , for μ a Γ- invariant state of C and $\varphi = \mu \circ P$, we shall prove that φ is central . Indeed , for $x = u_1 c_1$, $y = u_2 c_2$ with $u_1 , u_2 \in U$, $c_1 , c_2 \in C$, we have

$$\begin{aligned} \varphi(xy) &= \mu(P(u_1 c_1 u_2 c_2 u_1 u_1^*)) = \mu(u_1 P(c_1 u_2 c_2 u_1) u_1^*) = \\ &= \mu(P(c_1 u_2 c_2 u_1)) \quad = \mu(c_1 P(u_2 c_2 u_1)) \quad = \\ &= \mu(P(u_2 c_2 u_1 c_1)) \quad = \varphi(yx) \quad . \end{aligned}$$

This ends the proof since c.l.m.(UC) = A .

Q.E.D.

I.3.3. We shall prove that the representation π_μ is standard , more precisely we have :

PROPOSITION . *Let* μ *be a* Γ- *quasi-invariant probability measure on* Ω . *Then* ξ_μ *is cyclic and separating for* $\pi_\mu(A)''$.

Proof . By the GNS construction ξ_μ is cyclic , so all we have to prove is that ξ_μ is separating . For $x \in A$ we have

$$(\pi_\mu(x)\xi_\mu|\xi_\mu) = \varphi_\mu(x) = \mu(P(x)) = \int_\Omega P(x)(t)\, d\mu(t) \quad .$$

Suppose $x \in \pi_\mu(A)''$ is such that $x\xi_\mu = 0$. By the Kaplansky density theorem and the separability of A there is a norm-bounded sequence of elements $x_k \in A$ such that $\pi_\mu(x_k)$ converges strongly to x . Hence

$$(1) \qquad \lim_{k\to\infty}\int_\Omega P(x_k^* x_k)(t)\, d\mu(t) = \lim_{k\to\infty}\|\pi_\mu(x_k)\xi_\mu\|^2 = \|x\,\xi_\mu\|^2 = 0 \ .$$

To prove that $x = 0$ it will be enough to show that $x\eta = 0$ for η in a total subset of H_μ. Thus , ξ_μ being cyclic and $A =$ c.l.m.(UC) , it will be sufficient to prove that

$$x\,\pi_\mu(uc)\xi_\mu = 0 \qquad \text{for all} \quad u \in U \ , \ c \in C \ ,$$

that is

$$\lim_{k\to\infty}\|\pi_\mu(x_k uc)\xi_\mu\| = 0 \qquad \text{for all} \quad u \in U \ , \ c \in C \ ,$$

or equivalently

$$(2) \qquad \lim_{k\to\infty}\int_\Omega (c^*c)(t) \quad (u^*P(x_k^*x_k)u))(t)\, d\mu(t) = 0 \quad ; \ u \in U \ , \ c \in C \ .$$

Thus we must prove that (1) $\Longrightarrow$ (2) and , since μ is

Γ - quasi-invariant , this will follow from the following more general fact :

"If $\{f_k\}$ is a uniformly bounded sequence of positive measurable functions on Ω , $f \in L^{\infty}(\Omega, \mu)$ and ν a probability measure on Ω absolutely continuous with respect to μ , then

$$\lim_{k\to\infty}\int_{\Omega} f_k(t)\, d\mu(t) = 0 \Longrightarrow \lim_{k\to\infty}\int_{\Omega} f(t)f_k(t)\, d\nu(t) = 0"$$

To verify this assertion , let $h = \frac{d\nu}{d\mu}$ be the Radon-Nikodym derivative of ν with respect to μ . Then $h \in L^1(\Omega, \mu)$, $h \geqslant 0$ and we have to show that

$$\lim_{k\to\infty}\int_{\Omega} h(t)f(t)f_k(t)\, d\mu(t) = 0 \quad .$$

Consider

$$M = \sup\left\{\|f\| \quad , \|f_k\| \quad ; k \in \mathbb{N}\right\} \quad ,$$

$$E_n = \left\{t \in \Omega \ ; \ h(t) \leqslant n\right\} \quad .$$

Then

$$\lim_{n\to\infty}\int_{\Omega \setminus E_n} h(t)\, d\mu(t) = 0 \quad ,$$

so for given $\varepsilon > 0$ there is n_0 such that

$$\int_{\Omega \setminus E_{n_0}} h(t)\, d\mu(t) \leqslant \frac{\varepsilon}{2M^2} \quad .$$

On the other hand , by the hypothesis there is $k_\varepsilon \in \mathbb{N}$ such that , for $k \geqslant k_\varepsilon$,

$$\int_{\Omega} f_k(t)\, d\mu(t) \leqslant \frac{\varepsilon}{2n_0M} \quad .$$

Hence for $k \geqslant k_\varepsilon$ we have :

$$\int_{\Omega} h(t)f(t)f_k(t)\, d\mu(t) = \int_{E_{n_o}} h(t)f(t)f_k(t)\, d\mu(t) + $$

$$+ \int_{\Omega \setminus E_{n_o}} h(t)f(t)f_k(t)\, d\mu(t) \leqslant$$

$$\leqslant n_o M \int_{\Omega} f_k(t)\, d\mu(t) +$$

$$+ M^2 \int_{\Omega \setminus E_{n_o}} h(t)\, d\mu(t) \leqslant$$

$$\leqslant n_o M \frac{\varepsilon}{2n_o M} + M^2 \frac{\varepsilon}{2M^2} = \varepsilon .$$

This ends the proof of the Proposition .

Q.E.D.

I.3.4. PROPOSITION . Let μ be a Γ - quasi-invariant probability measure on Ω . Then we have

$$\ker \pi_{\mu} = J_{\mu} .$$

Proof . By Proposition 3.3. we have

$$x \in \ker \pi_{\mu} \Longleftrightarrow \pi_{\mu}(x)\xi_{\mu} = 0 \Longleftrightarrow \varphi_{\mu}(x^* x) = 0$$

$$\Longleftrightarrow \int_{\Omega} P(x^* x)(t)\, d\mu(t) = 0$$

and since $P(x^* x)$ is a continuous function on Ω it follows

$$x \in \ker \pi_{\mu} \Longleftrightarrow P(x^* x)(t) = 0 , (\forall)\ t \in \Omega_{\mu}$$

$$\Longleftrightarrow x \in J_{\mu} .$$

Q.E.D.

I.3.5. The Gelfand spectrum of $\pi_{\mu}(C)$ is Ω_{μ} . Hence , via Gelfand isomorphisms $C \simeq C(\Omega)$, $\pi_{\mu}(C) \simeq C(\Omega_{\mu})$, the restriction of π_μ to C corresponds to

$$C(\Omega) \ni c \longmapsto c \mid \Omega_\mu \in C(\Omega_\mu) \quad .$$

The state of $C(\Omega)$ corresponding to μ can be factored through this homomorphism , the corresponding state of $C(\Omega_\mu)$ being the restriction of the measure μ to its support Ω_μ . For any element $\pi_\mu(C) \ni \pi_\mu(c) \simeq f \in C(\Omega_\mu)$ we have

$$(\pi_\mu(c)\xi_\mu \mid \xi_\mu) = \int_{\Omega_\mu} f(t)\, d\mu(t) \quad .$$

Moreover , we know that the vector state $x \longmapsto (x\xi_\mu \mid \xi_\mu)$ is a faithful normal state of $\pi_\mu(C)''$.

It follows then from known results ([6], Prop. 1 , § 7 , Chap. I) that :

(3) The Gelfand isomorphism $\pi_\mu(C) \simeq C(\Omega_\mu)$ has a unique extension to a normal isomorphism

$$\pi_\mu(C)'' \simeq L^\infty(\Omega_\mu, \mu) \simeq L^\infty(\Omega, \mu)$$

such that , for $\pi_\mu(C)'' \ni c \simeq f \in L^\infty(\Omega, \mu)$, we have

$$(c\xi_\mu \mid \xi_\mu) = \int_\Omega f(t)\, d\mu(t) \quad .$$

The following result is also well known

(4) If τ is a faithful normal semifinite trace on $L^\infty(\Omega, \mu)$ then there is a unique sigma-finite positive measure ν on Ω , equivalent to μ , such that

$$\tau(f) = \int_\Omega f(t)\, d\nu(t)$$

for any $f \in L^\infty(\Omega, \mu)$, $f \geqslant 0$.

I.3.6. PROPOSITION . <u>Let</u> μ_1, μ_2 <u>be</u> Γ - <u>quasi-invariant probability measures on</u> Ω . <u>Then the representations</u> π_{μ_1}, π_{μ_2} <u>are unitarily equivalent if and only if the measures</u> μ_1, μ_2 <u>are equivalent</u> .

<u>Proof</u> . Since π_{μ_1} , π_{μ_2} are equivalent , they have the same kernel , so that , by Proposition I.3.4. , $\Omega_{\mu_1} = \Omega_{\mu_2}$. Moreover , there is a normal isomorphism $\pi_{\mu_1}(C)'' \simeq \pi_{\mu_2}(C)''$ which extends the isomorphism

$$\pi_{\mu_1}(C) \ni \pi_{\mu_1}(c) \simeq \pi_{\mu_2}(c) \in \pi_{\mu_2}(C)$$

That is , there is a normal isomorphism

$$L^{\infty}(\Omega_{\mu_1}, \mu_1) \simeq L^{\infty}(\Omega_{\mu_2}, \mu_2)$$

equal to the identity on $C(\Omega_{\mu_1}) = C(\Omega_{\mu_2})$. This easily yields the equivalence of μ_1 and μ_2 .

Conversely , suppose μ_1, μ_2 are equivalent . Consider

$$h = \frac{d\mu_2}{d\mu_1} \in L^1(\Omega, \mu_1) \quad , \quad h \geqslant 0 \quad .$$

Then there is a sequence $c_n \in C = C(\Omega)$, $c_n \geqslant 0$, such that $\{c_n{}^2\}$ converges in $L^1(\Omega, \mu_1)$ to h and therefore

a) $\{c_n\}$ is a Cauchy sequence in $L^2(\Omega, \mu_1)$;

b) $\int_{\Omega} f(t)h(t)\, d\mu_1(t) = \lim_{n\to\infty} \int_{\Omega} f(t)c_n(t)^2\, d\mu_1(t)$, $(\forall)\ f \in C(\Omega)$.

Next we have

$$\|\pi_{\mu_1}(c_n)\xi_{\mu_1} - \pi_{\mu_1}(c_m)\xi_{\mu_1}\|^2 = \int_{\Omega} |c_n(t) - c_m(t)|^2\, d\mu_1(t)$$

and from a) it follows that $\{\pi_{\mu_1}(c_n)\xi_{\mu_1}\}$ is a Cauchy sequence in H_1 . Put $\eta = \lim_{n\to\infty} \pi_{\mu_1}(c_n)\xi_{\mu_1} \in H_{\mu_1}$. In view of b) , for all $x \in A$ we have :

$$(\pi_{\mu_1}(x)\eta \mid \eta) = \lim_{n\to\infty} (\pi_{\mu_1}(c_n x c_n)\xi_{\mu_1} \mid \xi_{\mu_1}) =$$

$$= \lim_{n\to\infty} \int_\Omega P(x)(t) c_n(t)^2 \, d\mu_1(t) =$$

$$= \int_\Omega P(x)(t) h(t) \, d\mu_1(t) =$$

$$= \int_\Omega P(x)(t) \, d\mu_2(t) =$$

$$= (\pi_{\mu_2}(x)\xi_{\mu_2} \mid \xi_{\mu_2}) \quad ,$$

hence

$$\|\pi_{\mu_2}(x)\xi_{\mu_2}\| = \|\pi_{\mu_1}(x)\eta\| \quad .$$

Thus , there is a unique isometry V of H_{μ_2} into H_{μ_1} such that

$$V(\pi_{\mu_2}(x)\xi_{\mu_2}) = \pi_{\mu_1}(x)\eta \, , \quad x \in A \quad .$$

Clearly , V is intertwinning for π_{μ_2} and π_{μ_1} .

Since the same kind of argument shows that π_{μ_1} is also equivalent to a subrepresentation of π_{μ_2} , the Schröder - Bernstein type theorem gives us the desired result .

Q.E.D.

Let us emphasize that , the representations π_μ being standard , two of them are quasi-equivalent if and only if they are unitarily equivalent .

I.3.7. Let μ be a Γ- quasi-invariant probability measure on Ω . For each $n \geq 0$ there is a strongly continuous conditional expectation

$$P_n^{\mu} : \pi_{\mu}(A)'' \longrightarrow \pi_{\mu}(C_n)' \cap \pi_{\mu}(A)''$$

defined by

$$P_n^{\mu}(x) = \sum_{i \in I_n} \pi_{\mu}(q_i)\, x\, \pi_{\mu}(q_i) \quad ,$$

where $\{q_i\}_{i \in I_n}$ are the minimal projections of C_n . Clearly ,

$$P_n^{\mu}(\pi_{\mu}(x)) = \pi_{\mu}(P_n(x)) \quad , \quad x \in A$$

and , since $P \circ P_n = P$, we have

$$(P_n^{\mu}(\pi_{\mu}(x))\xi_{\mu} \mid \xi_{\mu}) = (\pi_{\mu}(x)\xi_{\mu} \mid \xi_{\mu}) \quad , \quad x \in A \quad .$$

By the strong continuity of P_n^{μ} we infer

(5) $$(P_n^{\mu}(x)\xi_{\mu} \mid \xi_{\mu}) = (x\xi_{\mu} \mid \xi_{\mu}) \quad , \quad x \in \pi_{\mu}(A)'' \; .$$

Moreover , for any $x \in A$ we have

$$\begin{aligned} \|P_n^{\mu}(\pi_{\mu}(x))\xi_{\mu}\|^2 &= (\pi_{\mu}(P_n(x)^* P_n(x))\xi_{\mu} \mid \xi_{\mu}) \\ &= (\mu \circ P)(P_n(x)^* P_n(x)) \\ &\leqslant (\mu \circ P)(P_n(x^* x)) \\ &= \mu(P(x^* x)) \\ &= \|\pi_{\mu}(x)\xi_{\mu}\|^2 \end{aligned}$$

and , again by the strong continuity of P_n^{μ} ,

(6) $$\|P_n^{\mu}(x)\xi_{\mu}\| \leqslant \|x\xi_{\mu}\| \quad , \quad x \in \pi_{\mu}(A)'' \; .$$

By Lemma I.2.3. , there is a projection of norm one

$$P^{\mu} : \pi_{\mu}(A) \longrightarrow \pi_{\mu}(C)$$

such that

(7) $$P^{\mu}(\pi_{\mu}(x)) = \pi_{\mu}(P(x)) \quad , \quad x \in A \quad .$$

Since

(8) $$P^{\mu}(\pi_{\mu}(x)) = P_n^{\mu}(\pi_{\mu}(x)) \quad , \quad x \in A_n \; ,$$

it follows from (6) that

$$\|P^{\mu}(\pi_{\mu}(x))\xi_{\mu}\| \leqslant \|\pi_{\mu}(x)\xi_{\mu}\| \quad , \quad x \in \bigcup_{n=0}^{\infty} A_n \ .$$

Hence , for any $T' \in \pi_{\mu}(A)'$, we have

$$\|P^{\mu}(\pi_{\mu}(x))T'\xi_{\mu}\| \leqslant \|T'\|\|\pi_{\mu}(x)\xi_{\mu}\| \quad , \quad x \in \bigcup_{n=0}^{\infty} A_n \ .$$

Since by I.3.3. $\pi_{\mu}(A)'\xi_{\mu}$ is dense in H_{μ} , the preceding results show that P^{μ} is strongly continuous on bounded subsets of $\pi_{\mu}(\bigcup_{n=0}^{\infty} A_n)$. Using the Kaplansky density theorem , we can extend P^{μ} up to a linear map

$$P^{\mu} : \pi_{\mu}(A)'' \longrightarrow \pi_{\mu}(C)''$$

strongly continuous on bounded subsets . It follows that P^{μ} is a projection of norm one and also a normal map . Thus we have (see also [6], Th. 2 , § 4 , Ch.I) :

(9) P^{μ} <u>is a ultraweakly and ultrastrongly continuous conditional expectation of $\pi_{\mu}(A)''$ with respect to $\pi_{\mu}(C)''$</u>.

Owing to the relation (8) and to the continuity of P^{μ} and P^{μ}_n , it follows that

(10) $$P^{\mu}(x)\xi_{\mu} = \lim_{n\to\infty} P^{\mu}_n(x)\xi_{\mu} \quad , \quad x \in \pi_{\mu}(A)'' \ .$$

Then we have also

(11) $$(x\xi_{\mu}|\xi_{\mu}) = \int_{\Omega} P^{\mu}(x)(t)\, d\mu(t) \quad , \quad x \in \pi_{\mu}(A)'' \ ,$$

where $P^{\mu}(x)$ is regarded as an element of $L^{\infty}(\Omega, \mu)$. In particular ,

(12) <u>The conditional expectation</u> P^{μ} <u>is faithful</u> .

Also , clearly , we have

(13) $$P^{\mu}(u^*xu) = u^*P^{\mu}(x)u \quad , \quad x \in \pi_{\mu}(A)'' \ , \quad u \in \pi_{\mu}(U) \ .$$

I.3.8. PROPOSITION . <u>Let</u> μ <u>be a</u> Γ- <u>quasi-invariant probability measure on</u> Ω . <u>Then</u> $\pi_\mu(C)''$ <u>is a m.a.s.a. in</u> $\pi_\mu(A)''$.

<u>Proof</u> . Consider $x \in \pi_\mu(A)'' \cap \pi_\mu(C)'$. Since $x \in \pi_\mu(C_n)'$ we have $P_n^\mu(x) = x$, for each $n \geq o$. By (10) we infer

$$P^\mu(x)\xi_\mu = x\xi_\mu$$

and , since ξ_μ is separating , $x = P^\mu(x) \in \pi_\mu(C)''$.

Q.E.D.

I.3.9. PROPOSITION . <u>Let</u> μ <u>be a</u> Γ - <u>quasi-invariant probability measure on</u> Ω . <u>Then</u> π_μ <u>is a factor representation if and only if</u> μ <u>is</u> Γ - <u>ergodic</u> .

<u>Proof</u> . Suppose π_μ is a factor representation and let $p \in \pi_\mu(C)'' \simeq L^\infty(\Omega, \mu)$ be a Γ - invariant projection . Then $p \in \pi_\mu(A)'' \cap (\pi_\mu(UC))' = \pi_\mu(A)'' \cap \pi_\mu(A)'$ and hence is either 0 or 1 .

Conversely , suppose μ is Γ- ergodic and consider a central projection $p \in \pi_\mu(A)''$. By I.3.8., $p \in \pi_\mu(C)'' \simeq L^\infty(\Omega, \mu)$ and clearly p is Γ - invariant . Thus , p is either 0 or 1 .

Q.E.D.

I.3.10. PROPOSITION . <u>Let</u> μ <u>be a</u> Γ- <u>quasi-invariant probability measure on</u> Ω . <u>The representation</u> π_μ <u>is finite if and only if</u> μ <u>is equivalent to some</u> Γ - <u>invariant probability measure on</u> Ω . <u>Moreover , every finite representation of</u> A <u>is quasi-equivalent to a representation</u> π_μ .

Proof . Suppose π is a finite representation of A and let τ be a normal faithful finite trace on $\pi(A)''$ with $\tau(1) = 1$. Then the representation π is quasi-equivalent to the representation of A obtained via the GNS construction for the state $\tau \circ \pi$. Because $\tau \circ \pi$ is central , by Proposition I.3.2. there is a Γ- invariant probability measure ν on Ω such that $\tau \circ \pi = \nu \circ P$. If π is some π_μ , the equivalence of the measures μ and ν follows from Proposition I.3.6.

Conversely , if ν is a Γ- invariant probability measure on Ω , equivalent to μ , then π_μ and π_ν are equivalent by I.3.6. Moreover , φ_ν being central , ξ_ν is a trace-vector for the von Neumann algebra $\pi_\nu(A)''$.

Q.E.D.

I.3.11. PROPOSITION . Let μ be a Γ- quasi-invariant probability measure on Ω . The representation π_μ is semi-finite if and only if μ is equivalent to some sigma-finite Γ- invariant positive measure on Ω .

Proof . Let τ be a normal semifinite faithful trace on $\pi_\mu(A)''$. We shall prove that the restriction of τ to $\pi_\mu(C)''$ is semifinite . Thus , for any

$$y \in \pi_\mu(C)'' \quad \text{with} \quad y \geq 0 \ , \quad y \neq 0 \ ,$$

we must prove the existence of

$$z \in \pi_\mu(C)'' \quad \text{with} \quad 0 \leq z \leq y \quad \text{and} \quad 0 < \tau(z) < +\infty \ .$$

Since τ is semifinite and faithful , there is

$$x \in \Pi_\mu(A)'' \quad \text{with} \quad 0 \leqslant x \leqslant y \quad \text{and} \quad 0 < \tau(x) < +\infty \quad .$$

Moreover , since ξ_μ is a separating vector , we have $x^{1/2}\xi_\mu \neq 0$ and therefore

$$(x\xi_\mu | \xi_\mu) \neq 0 \quad .$$

Consider

$$x_n = P_n^\mu(x) \in \Pi_\mu(C_n)' \quad .$$

Because of the well known properties of the trace ([6], Prop. 1 , § 6 , Chap. I) , we have

$$\tau(x_n) = \tau(\sum_{i \in I_n} \Pi_\mu(q_i)\, x\, \Pi_\mu(q_i)) = \tau(\sum_{i \in I_n} x\Pi_\mu(q_i)) = \tau(x) \quad .$$

By the relation (5) from I.3.7. , we get

$$(x_n\xi_\mu | \xi_\mu) = (x\xi_\mu | \xi_\mu) \neq 0 \quad .$$

Note also that

$$0 \leqslant x_n \leqslant y \quad .$$

Thus , there exists a weak cluster point z of the sequence $\{x_n\}$ and we have

$$0 \leqslant z \leqslant y \quad ,$$

$$(z\xi_\mu | \xi_\mu) = (x\xi_\mu | \xi_\mu) \neq 0 \quad ,$$

$$z \in \bigcap_{n=0}^{\infty} \Pi_\mu(C_n)' = \Pi_\mu(C)'' \quad ,$$

$$\tau(z) \leqslant \liminf_{n \to \infty} \tau(x_n) = \tau(x) < +\infty \quad ,$$

where we have used the fact that $\Pi_\mu(C)''$ is a m.a.s.a. (I.3.8.) and the weak lower semi-continuity of τ .

The existence of a Γ- invariant sigma-finite positive measure ν on Ω equivalent to μ is now an easy consequence of the assertion (4) from I.3.5. applied to the restriction of τ to $\pi_\mu(C)'' \simeq L^\infty(\Omega, \mu)$.

Conversely , let ν be a sigma-finite Γ - invariant positive measure on Ω , equivalent to μ . For $x \in \pi_\mu(A)''$, $x \geq 0$, we define

(14) $$\tau(x) = \int_\Omega P^\mu(x)(t)\, d\nu(t)$$

which is correctly defined since ν is absolutely continuous with respect to μ . Thus we get a <u>weight</u> τ on $(\pi_\mu(A)'')^+$. That τ is <u>faithful</u> and <u>normal</u> follows from the corresponding properties of P^μ (see I.3.7., (9) and (12)) , μ being absolutely continuous with respect to ν . Also , that τ is <u>semifinite</u> is a direct consequence of the fact that ν is sigma-finite .

To prove that τ is a <u>trace</u> , we must show that

$$\tau(x^*x) = \tau(xx^*) \quad , \quad x \in \pi_\mu(A)'' \quad .$$

Consider $e_j \uparrow 1$ projections in $\pi_\mu(C)''$ with $\tau(e_j) < +\infty$. Then

$$e_j P^\mu(x^*e_jx) \uparrow P^\mu(x^*x) \quad \text{and} \quad e_j P^\mu(xe_jx^*) \uparrow P^\mu(xx^*)$$

and therefore

$$\tau(x^*x) = \lim_{j\to\infty} \int_\Omega e_j P^\mu(x^*e_jx)\, d\nu \quad ,$$

$$\tau(xx^*) = \lim_{j\to\infty} \int_\Omega e_j P^\mu(xe_jx^*)\, d\nu \quad .$$

Hence it will be sufficient to prove that

(15) $$\int_\Omega e_j P^\mu(xe_jy)\, d\nu = \int_\Omega e_j P^\mu(ye_jx)\, d\nu \quad ,$$

for all $x , y \in \pi_\mu(A)''$ and all j .

Let us now show that it is sufficient to prove (15) only

for $x , y \in \pi_\mu(A)$. Indeed , for $x , y \in \pi_\mu(A)''$ we can find , by Kaplansky's density theorem , two sequences $\{x_k\}$, $\{y_k\} \subset \pi_\mu(A)$ strongly convergent to x , y respectively . Since P^μ is strongly continuous on bounded sets and since $\nu_j = e_j \, d\nu$ are normal positive functionals on $\pi_\mu(C)'' \simeq L^\infty(\Omega, \mu)$, we have

$$\lim_{k\to\infty} \nu_j(P^\mu(x_k e_j y_k)) = \nu_j(P^\mu(x e_j y)) \quad ,$$

$$\lim_{k\to\infty} \nu_j(P^\mu(y_k e_j x_k)) = \nu_j(P^\mu(y e_j x)) \quad ,$$

that is ,

$$\lim_{k\to\infty} \int_\Omega e_j \; P^\mu(x_k e_j y_k) \, d\nu = \int_\Omega e_j \; P^\mu(x e_j y) \, d\nu \quad ,$$

$$\lim_{k\to\infty} \int_\Omega e_j \; P^\mu(y_k e_j x_k) \, d\nu = \int_\Omega e_j \; P^\mu(y e_j x) \, d\nu \quad .$$

Now we may further restrict the proof of (15) to the case when $x = u_1 c_1$, $y = u_2 c_2$ with $u_1 , u_2 \in \pi_\mu(U)$, $c_1 , c_2 \in \pi_\mu(C)$. But then , by the Γ - invariance of ν , we have

$$\begin{aligned} \int_\Omega e_j \; P^\mu(x e_j y) \, d\nu &= \int_\Omega u_1 \; P^\mu(c_1 e_j u_2 c_2 e_j u_1) \; u_1^* \, d\nu \\ &= \int_\Omega P^\mu(c_1 e_j u_2 c_2 e_j u_1) \, d\nu \\ &= \int_\Omega e_j \; P^\mu(u_2 c_2 e_j u_1 c_1) \, d\nu \\ &= \int_\Omega e_j \; P^\mu(y e_j x) \, d\nu \quad . \end{aligned}$$

Q.E.D.

The preceding Proposition shows that all semifinite representations π_μ can be obtained by choosing , for any given sigma-finite Γ- invariant positive measure ν , an equivalent probability measure μ . Then the trace on $\pi_\mu(A)''$ is given by formula (14) .

Note also that the above proof shows that the restriction of the trace to $\pi_\mu(C)''$ is semifinite . It follows that , in the factor case , π_μ is discrete (i.e. type I) if and only if ν , or equivalently μ , is discrete (i.e. completely atomic) .

I.3.12. Summing up the results of the preceding Sections , we obtain the following Theorem .

THEOREM . Consider an AF - algebra A with m.a.s.a. C , conditional expectation $P : A \longrightarrow C$ and associated topological dynamical system (Ω, Γ) . Let μ be a Γ- quasi-invariant probability measure on Ω and π_μ be the representation of A with cyclic vector ξ_μ associated by the GNS construction to the state $\mu \circ P$ of A . Then :

1) π_μ is standard , i.e. ξ_μ is (cyclic and) separating for the von Neumann algebra $\pi_\mu(A)''$.

2) $\pi_\mu(C)''$ is a m.a.s.a. in $\pi_\mu(A)''$.

3) $\ker \pi_\mu = \{x \in A \; ; \; P(x^*x)(t) = 0 \; , \; (\forall) \; t \in \text{support of } \mu\}$.

4) $\pi_\mu(A)''$ is a factor $\Longleftrightarrow$ μ is Γ- ergodic .

5) $\pi_\mu(A)''$ is finite $\Longleftrightarrow$ μ is equivalent to a Γ- invariant probability measure on Ω . Moreover , every finite representation of A is quasi-equivalent to some π_μ.

6) $\pi_\mu(A)''$ <u>is semifinite</u> $\Longleftrightarrow$ μ <u>is equivalent to a</u> Γ - <u>invariant sigma-finite positive measure on</u> Ω.

7) π_{μ_1}, π_{μ_2} <u>are equivalent</u> $\Longleftrightarrow$ μ_1, μ_2 <u>are equivalent</u>.

<u>Moreover</u>, <u>concerning the type of</u> π_μ, <u>in the factor case</u> (<u>i.e.</u> <u>when</u> μ <u>is ergodic</u>) <u>we have</u> :

(i) π_μ <u>is of type</u> I_n $\Longleftrightarrow$ μ <u>is discrete</u> <u>and</u> $\text{card}(\text{supp}\,\mu) = n$; $n \in \mathbb{N}$.

(ii) π_μ <u>is of type</u> I_∞ $\Longleftrightarrow$ μ <u>is discrete</u> <u>and</u> $\text{card}(\text{supp}\,\mu) = \infty$.

(iii) π_μ <u>is of type</u> II_1 $\Longleftrightarrow$ μ <u>is equivalent to a non-discrete</u> Γ- <u>invariant probability measure on</u> Ω .

(iv) π_μ <u>is of type</u> II_∞ $\Longleftrightarrow$ μ <u>is equivalent to a non-discrete</u> Γ- <u>invariant sigma-finite infinite positive measure on</u> Ω .

(v) π_μ <u>is of type</u> III $\Longleftrightarrow$ μ <u>is not equivalent to any sigma-finite</u> Γ- <u>invariant positive measure on</u> Ω .

In the last case , i.e. if μ is not equivalent to any sigma-finite Γ - invariant positive measure on Ω , then the group Γ is said to be <u>non-measurable with respect to</u> μ ([ ]).

I.3.13. R.Powers ([24], 2.5.) has given a necessary and sufficient condition that a state of an UHF - algebra be a factor state . This result was extended by O.Bratteli ([1], 4.4.) to the case of AF - algebras :

(16) Let $A = \langle \bigcup_{n=0}^{\infty} A_n \rangle$ be an AF - algebra and φ a state of A. The representation π_φ associated to φ by the GNS construction is a factor representation if and only if for every $x \in A$ there is $n \geqslant 0$ such that

$$|\varphi(xy) - \varphi(x)\varphi(y)| \leqslant \|\pi_\varphi(y)\|$$

for all $y \in A_n'$.

The same way, R.Powers ([24], 2.7.) for UHF - algebras and O.Bratteli ([1], 4.5.) for AF - algebras, have shown:

(17) Let $A = \langle \bigcup_{n=0}^{\infty} A_n \rangle$ be an AF - algebra and φ_1, φ_2 states of A. The representations π_{φ_1}, π_{φ_2} associated to φ_1, φ_2 by the GNS construction, are quasi-equivalent if and only if for every $\varepsilon > 0$ there is $n \geqslant 0$ such that

$$|\varphi_1(x) - \varphi_2(x)| \leqslant \varepsilon \|\pi_{\varphi_1}(x)\|$$

for all $x \in A_n'$.

We shall use these results in the case of states $\mu \circ P$ in order to obtain ergodicity and equivalence criteria for Γ- quasi-invariant probability measures on Ω. By Γ_n we shall denote the image of U_n in Γ.

I.3.14. PROPOSITION. *Let μ be a Γ - quasi-invariant probability measure on Ω. Then μ is Γ - ergodic if and only if for every $f \in C(\Omega)$ there is $n \geqslant 0$ such that for all Γ_n- invariant $g \in C(\Omega)$ we have*

$$\left| \int_\Omega fg \, d\mu - \left(\int_\Omega f \, d\mu\right)\left(\int_\Omega g \, d\mu\right) \right| \leqslant \|g\|_{L^\infty(\Omega,\mu)} \quad .$$

Proof . As we have seen , μ is Γ - ergodic if and only if π_μ is a factor representation . Hence , by (16) , μ is Γ - ergodic if and only if the following holds :

$$(18) \quad \left[\begin{array}{l} \text{for every } x \in A \text{ there is } n \geqslant 0 \text{ such that} \\ \qquad \left| \mu(P(xy)) - \mu(P(x))\,\mu(P(y)) \right| \leqslant \left\| \pi_\mu(y) \right\| \\ \text{for all } y \in A'_n \end{array} \right.$$

We must prove that this is equivalent to

$$(19) \quad \left[\begin{array}{l} \text{for every } x \in C \text{ there is } n \geqslant 0 \text{ such that} \\ \qquad \left| \mu(xy) - \mu(x)\,\mu(y) \right| \leqslant \left\| \pi_\mu(y) \right\| \\ \text{for all } \Gamma_n \text{ - invariant } y \in C \, . \end{array} \right.$$

That (18) $\Longrightarrow$ (19) follows immediately from the fact that $A'_n \cap C$ is the set of all Γ_n - invariant elements of C .

Conversely , suppose (19) holds . Clearly it is sufficient to establish that (18) holds for x in the dense set $\bigcup_{n=0}^{\infty} A_n$. Thus , for $x \in A_m$ choose $n > m$ such that for $z \in A'_n \cap C$ we have

$$\left| \mu(P(x)z) - \mu(P(x))\,\mu(z) \right| \leqslant \left\| \pi_\mu(z) \right\| \; .$$

For $y \in A'_n \subset A'_m$ we have $P(y) \in A'_n \cap C$ (by I.1.5.) and $P(xy) = P(x)P(y)$ (by I.1.7.) . It follows that

$$\left| \mu(P(xy)) - \mu(P(x))\,\mu(P(y)) \right| =$$

$$= \left| \mu(P(x)P(y)) - \mu(P(x))\,\mu(P(y)) \right| \leqslant$$

$$\leqslant \left\| \pi_\mu(P(y)) \right\| =$$

$$= \left\| P^{\mu}(\pi_{\mu}(y)) \right\| \leq$$

$$\leq \left\| \pi_{\mu}(y) \right\| \quad .$$

Q.E.D.

I.3.15. From the other result , (17) , of R. Powers and O. Bratteli we obtain in a similar way :

PROPOSITION . <u>Let</u> μ_1 , μ_2 <u>be</u> Γ- <u>quasi-invariant probability measures on</u> Ω . <u>Then</u> μ_1 <u>and</u> μ_2 <u>are equivalent if and only if , for every</u> $\varepsilon > 0$ <u>there is</u> $n \geq 0$ <u>such that</u>

$$\left| \int_{\Omega} f \, d\mu_1 - \int_{\Omega} f \, d\mu_2 \right| \leq \varepsilon \|f\|_{L^{\infty}(\Omega,\mu_1)}$$

<u>for all</u> Γ_n - <u>invariant</u> $f \in C(\Omega)$.

I.3.16. <u>The irreducible representations</u> ρ_{μ} .

Let μ be a Γ- quasi-invariant measure on Ω . Consider, for every $f \in C(\Omega)$, the operator $\rho_{\mu}(T_f)$ on $L^2(\Omega,\mu)$ defined by

$$\rho_{\mu}(T_f)\, h = fh \qquad ; h \in L^2(\Omega,\mu)$$

and , for every $\gamma \in \Gamma$, the operator $\rho_{\mu}(V_{\gamma})$ on $L^2(\Omega,\mu)$ defined by

$$(\rho_{\mu}(V_{\gamma})\, h)\,(t) = \left(\frac{d\mu^{\gamma}}{d\mu}\right)^{1/2}(t)\; h(\gamma^{-1}(t)) \ , \ t \in \Omega ; \ h \in L^2(\Omega,\mu).$$

Then $T_f \longrightarrow \rho_{\mu}(T_f)$, $V_{\gamma} \longrightarrow \rho_{\mu}(V_{\gamma})$ can be extended to a * - representation ρ_{μ} of the C^*- algebra $A(\Omega,\Gamma)$, as it is easily seen .

Composing this * - representation of $A(\Omega,\Gamma)$ with the

$*$ - isomorphism of A onto $A(\Omega,\Gamma)$ (see I.1.10.) , we get a $*$ - representation of A , which will be still denoted by ρ_μ . This notation makes sense only in the presence of a fixed $*$ - isomorphism of A onto $A(\Omega,\Gamma)$ (that is , a fixed system of matrix units for A) , since different $*$ - isomorphisms (that is , different systems of matrix units) may yield different $*$ - representations of A .

Concerning the irreducibility of ρ_μ we have :

ρ_μ <u>is irreducible</u> $\Longleftrightarrow$ μ <u>is</u> Γ - <u>ergodic</u>

Indeed , $L^\infty(\Omega,\mu)$ is a m.a.s.a. in $L(L^2(\Omega,\mu))$ and the commutant of $\rho_\mu(A)$ is the set of Γ- invariant elements of $L^\infty(\Omega,\mu)$.

Let us also remark that $\rho_\mu | C$ does not depend on the choice of the system of matrix units which implements the $*$ - isomorphism between A and $A(\Omega,\Gamma)$ and therefore the equivalence of ρ_{μ_1} and ρ_{μ_2} , even obtained from different $*$ - isomorphisms , entails the equivalence of the measures μ_1 and μ_2 .

Conversely , if μ_1 and μ_2 are equivalent , then the multiplication by $(d\mu_1/d\mu_2)^{1/2}$ implements an unitary equivalence between ρ_{μ_1} and ρ_{μ_2} regarded as representations of $A(\Omega,\Gamma)$ and therefore , between the corresponding representations of A , obtained via the same $*$ - isomorphism $A \simeq A(\Omega,\Gamma)$.

I.3.17. In this Section a general method of constructing $*$- representations of $A(\Omega,\Gamma)$ and hence also of A , will be outlined . Unfortunately , it is of limited use for our purposes since it is difficult to decide when such a representation is a factor representation .

For $f \in C(\Omega)$ and $\gamma \in \Gamma$ let us denote by $f^{\gamma} \in C(\Omega)$ the function defined by

$$f^{\gamma}(t) = f(\gamma^{-1}(t)) \quad , \quad t \in \Omega \quad .$$

Then , if ρ is a representation of $A(\Omega, \Gamma)$ on some Hilbert space H , we have that $\gamma \longmapsto \rho(V_{\gamma})$ is a unitary representation of Γ and $f \longmapsto \rho(T_f)$ is a $*$ - representation of $C(\Omega)$. Moreover , we have

$$\rho(V_{\gamma})\,\rho(T_f)\,\rho(V_{\gamma^{-1}}) = (T_{f^{\gamma}}) \quad .$$

This is what is known as a <u>covariant</u> <u>representation</u> (see for example [9], Def. 2 ;[31], 3.1.) of the dynamical system (Ω, Γ) and , in case H is separable , the following is known (see for example [16], Prop. 3.5.) :

To give a representation ρ of the dynamical system (Ω, Γ) is equivalent to give :

(i) a Γ - quasi-invariant measure μ on Ω (or rather an equivalence class of such measures);

(ii) a μ - measurable field of Hilbert spaces $t \longmapsto H_t$ over Ω ;

(iii) for each $\gamma \in \Gamma$, a measurable field of Hilbert space isomorphisms $\Psi_{\gamma,t} : H_{\gamma(t)} \longrightarrow H_t$;

such that

$$(*) \quad \Psi_{\gamma_1\gamma_2,t} = \Psi_{\gamma_2,t} \circ \Psi_{\gamma_1,\gamma_2(t)} \quad , \quad \Psi_{\varepsilon,t} = I \quad .$$

Then $\rho(T_f)$ is the multiplication operator by f and

$$(\rho(V_\gamma)\eta)(t) = \left(\frac{d\mu^\gamma}{d\mu}\right)^{1/2} \Psi_{\gamma^{-1},t}\,\eta(\gamma^{-1}(t))$$

for any $\eta \in \int_\Omega^\oplus H_t \, d\mu(t)$.

Let us remark that :

In order that the preceding representation of the dynamical system should yield a representation of $A(\Omega, \Gamma)$, it is necessary and sufficient that the following additional requirement be satisfied :

(**) $\Psi_{\gamma,t} = \Psi_{\gamma',t}$ for all $t \in \Omega$ such that $\gamma(t) = \gamma'(t)$.

It is also easily seen that

A necessary condition for the factoriality of the above representation is the ergodicity of μ and the requirement that $\dim H_t$ be almost everywhere constant .

For instance , in the case of the one dimensional trivial field of Hilbert spaces over Ω and $\Psi_{\gamma,t} = I$ for all $\gamma \in \Gamma$ and $t \in \Omega$, we get the representations ρ_μ of I.3.16.

Further , in the above general context ,

The equivalence of two such factor representations entails the equivalence of the corresponding measures on Ω and the equality of the numbers $\dim H_t$.

Moreover , in view of the special nature of our group Γ , an infinite algorithm can always be given for finding the solutions of $(*)$ and $(**)$, as in the case of the canonical anticommutation relations ([12]; see also [14]).

I.3.18. Most of what has been presented in this Chapter has its roots in the study of the representations of the canonical anticommutation relations and of the associated UHF - algebra and topological dynamical system. In this case Γ can be replaced by a smaller group Γ_0 freely acting on Ω and which, in the presence of a measure μ on Ω, has the same "full group" as Γ. Then the C^*- algebra $A(\Omega,\Gamma_0)$ is isomorphic to the cross-product of $C(\Omega)$ by Γ_0 and results similar to Theorem I.3.12. are well known.

The representations π_μ we have considered correspond in fact, via the isomorphism $A \simeq A(\Omega,\Gamma)$, to the "cross-product construction" of W. Krieger ([22]) for the dynamical systems (Ω,μ,Γ).

Namely, given an arbitrary dynamical system (Ω,μ,Γ), even if Γ does not act freely, W. Krieger has constructed a standard von Neumann algebra $\mathcal{A}(\mu)$ together with a m.a.s.a. $\mathcal{A}_0(\mu)$ in $\mathcal{A}(\mu)$ and has described the type of $\mathcal{A}(\mu)$ in a manner completely similar to that in Theorem I.3.12. A detailed exposition of W. Krieger's construction can be found in the book of A. Guichardet ([17],Chap. VII), where it is also pointed out that there is a unique conditional expectation of $\mathcal{A}(\mu)$ with respect to $\mathcal{A}_0(\mu)$.

The construction of W. Krieger shows that $\mathcal{A}(\mu)$ is generated by a covariant representation of the dynamical system (Ω, Γ). This extends to a $*$ - representation of $A(\Omega, \Gamma)$ and therefore, via the $*$ - isomorphism $A \simeq A(\Omega, \Gamma)$, to a $*$ - representation of A. It can be shown that this representation is unitarily equivalent to the representation π_μ in such a way that $\mathcal{A}_0(\mu)$ corresponds to $\pi_\mu(C)''$, the conditional expectation of $\mathcal{A}(\mu)$ with respect to $\mathcal{A}_0(\mu)$ corresponds to P^μ and the state of A associated to a certain cyclic separating vector for $\mathcal{A}(\mu)$ pointed out by W. Krieger corresponds to $\mu \circ P$.

Our choice of an exposition where W. Krieger's construction does not explicitely appear was motivated by the fact that once Ω and P are fixed, the representation does not depend on the isomorphism chosen between A and $A(\Omega, \Gamma)$ (that is, on the systems of matrix units). Also, to make our exposition more self-contained, we had to reprove in this frame-work some known results in the case of W. Krieger's construction.

CHAPTER II THE C^*- ALGEBRA ASSOCIATED TO A DIRECT LIMIT OF COMPACT GROUPS

§ 1 The L - algebra associated to a direct limit of compact groups

Let us consider a sequence

$$\{e\} = G_0 \subset G_1 \subset \dots \subset G_n \subset G_{n+1} \subset \dots$$

of separable compact groups such that each G_n is a closed subgroup in G_{n+1} of Haar measure zero . Let further G_∞ denote the direct limit of the groups G_n , endowed with the direct limit topology .

II.1.1. The measure algebras $M(G_n)$ define a direct limit of involutive Banach algebras , the completion of which is an involutive Banach algebra we shall denote by M . The group algebra $L^1(G_n)$ is a closed ideal in $M(G_n)$, hence

$$L_{(n)} = \sum_{k=1}^{n} L^1(G_k)$$

is an involutive subalgebra of $M(G_n)$.

Consider the involutive Banach subalgebra $L = L(G_\infty)$ of M defined by

$$L = L(G_\infty) = \overline{\bigcup_{n=1}^{\infty} L_{(n)}} \subset M \quad .$$

The measure-theoretic assumption made at the beginning insures that for $\mu_k \in L^1(G_k)$, $k = 1, \dots, n,$ we have

$$\left\| \sum_{k=1}^{n} \mu_k \right\|_{L_{(n)}} = \sum_{k=1}^{n} \|\mu_k\|_{L^1(G_k)} \quad .$$

Hence $L_{(n)}$ is closed in M and it is the topological direct sum of the $L^1(G_k)$'s , $k = 1,\ldots,n$. It follows that <u>each element</u> $\mu \in L$ <u>can be uniquely represented by a series</u>

$$\mu = \sum_{k=1}^{\infty} \mu_k \ , \quad \mu_k \in L^1(G_k)$$

<u>and</u>

$$\|\mu\|_L = \sum_{k=1}^{\infty} \|\mu_k\|_{L^1(G_k)}$$

For further use we also consider

$$L^{(n)} = \overline{\sum_{k>n} L^1(G_k)}$$

Then $L^{(n)}$ is a closed two-sided ideal of L and the quotient algebra $L/L^{(n)}$ is isomorphic to $L_{(n)}$.

The algebra $L = L(G_\infty)$ can be used in the study of the factor representations of G_∞ . In fact, as we shall see, the factor representations of L correspond to factor representations of G_∞ or of some G_n .

II.1.2. Let ρ be a continuous unitary representation of G_∞ . Then we can associate a representation π_ρ of L as follows. The restrictions ρ_n of ρ to G_n define representations of the measure algebras $M(G_n)$. These yield a representation of M . Finally, restricting this representation to L , we get the representation π_ρ we were looking for.

For completness we must also record a second kind of representations of L . For ρ_n a continuous unitary representation of G_n we get a representation of $M(G_n)$ and, by restriction, a representation of $L_{(n)}$. Since $L_{(n)}$ is isomorphic to $L/L^{(n)}$, this yields a representation π_{ρ_n} of L .

II.1.3. Before going any further let us fix for each n a sequence

$$\{u_j^{(n)}\}_{j\in\mathbb{N}} \subset L^1(G_n)$$

such that

$$u_j^{(n)} \geqslant 0$$

$$\|u_j^{(n)}\|_{L^1(G_n)} = 1$$

$$\text{supp } u_j^{(n)} \downarrow \{e\}$$

Then $\{u_j^{(n)}\}_{j\in\mathbb{N}}$ is an approximate unit for $L^{(n-1)}$, i.e. :

$$\lim_{j\to\infty} \|u^{(n)} * \varphi - \varphi\| = 0 \quad , \quad \varphi \in L^{(n-1)} \quad .$$

We shall use repeatedly the following remark. If ρ is a continuous unitary representation of a compact group and τ is its extension to the measure group algebra and if a sequence μ_n of measures converges weakly* to a measure μ , then $\tau(\mu_n)$ converges in the weak operator topology to $\tau(\mu)$.

II.1.4. Now let π be a factor representation of L . There are two cases :

a). <u>Suppose there is</u> $n \in \mathbb{N}$ <u>such that</u> $\pi(L^1(G_{n+1})) = 0$. Since $L^1(G_{n+1})$ contains an approximate unit for $L^{(n)}$, we have $\pi(L^{(n)}) = 0$. Let $n_o \in \mathbb{N}$ be the smallest $n \in \mathbb{N}$ such that $\pi(L^{(n)}) = 0$. We may view π as a representation of

$L_{(n_o)} \simeq L / L^{(n_o)}$. Since $L^1(G_{n_o})$ is an ideal of $L_{(n_o)}$ and since π is factorial, it follows that $\pi(u_j^{(n_o)})$ converges in the strong operator topology to the identity operator. Let ρ_{n_o} be the representation of G_{n_o} corresponding to $\pi | L^1(G_{n_o})$ and τ_{n_o} be its extension to $M(G_{n_o})$. For any $\mu \in L_{(n_o)}$ we have

$$\begin{aligned} \tau_{n_o}(\mu) &= \text{wo} - \lim_{j \to \infty} \tau_{n_o}(\mu * u_j^{(n_o)}) \\ &= \text{wo} - \lim_{j \to \infty} \pi(\mu * u_j^{(n_o)}) \\ &= \text{wo} - \lim_{j \to \infty} \pi(\mu)\pi(u_j^{(n_o)}) = \pi(\mu) \ . \end{aligned}$$

This means that

$$\pi = \pi_{\rho_{n_o}}$$

b). <u>Suppose</u> $\pi(L^1(G_n)) \neq 0$ <u>for</u> <u>all</u> $n \in \mathbb{N}$.

Since $L^{(n-1)}$ is an ideal of L and since π is factorial, it follows that $\pi(u_j^{(n)})$ converges in the strong operator topology to the identity operator and so $\pi | L^1(G_n)$ is non degenerate. Let ρ_n , τ_n be the representations of G_n , $M(G_n)$ respectively , corresponding to $\pi | L^1(G_n)$. We shall prove that

$$(1) \qquad \tau_{n+1} | M(G_n) = \tau_n \ .$$

Clearly , this implies

$$\rho_{n+1} | G_n = \rho_n \ ,$$

which allows us to define a continuous representation ρ of G_∞ such that

$$\pi = \pi_\rho \quad .$$

The following computation , with $\mu \in M(G_n)$, establishes (1) :

$$\begin{aligned} \tau_n(\mu) &= \text{wo-}\lim_{j\to\infty} \tau_n(\mu * u_j^{(n)}) \\ &= \text{wo-}\lim_{j\to\infty} \pi(\mu * u_j^{(n)}) \\ &= \text{wo-}\lim_{j\to\infty} \left(\text{wo-}\lim_{i\to\infty} \pi(\mu * u_j^{(n)})\pi(u_i^{(n+1)})\right) \\ &= \text{wo-}\lim_{j\to\infty} \left(\text{wo-}\lim_{i\to\infty} \pi(\mu * u_j^{(n)} * u_i^{(n+1)})\right) \\ &= \text{wo-}\lim_{j\to\infty} \left(\text{wo-}\lim_{i\to\infty} \tau_{n+1}(\mu * u_j^{(n)} * u_i^{(n+1)})\right) \\ &= \text{wo-}\lim_{j\to\infty} \tau_{n+1}(\mu * u_j^{(n)}) \\ &= \tau_{n+1}(\mu) \quad . \end{aligned}$$

II.1.5. Summing up the preceding discussion , we obtain the following

THEOREM. <u>A factor representation of</u> $L(G_\infty)$ <u>is always a representation</u> π_{ρ_n} <u>where</u> ρ_n <u>is a factor representation of some</u> G_n , $n \in \mathbb{N} \cup \{\infty\}$.

<u>Thus , there is a canonical one-to-one correspondence between the factor representations of</u> $L(G_\infty)$ <u>and the disjoint union of the factor representations of the</u> G_n<u>'s</u> $(n \in \mathbb{N} \cup \{\infty\})$.

It is clear that <u>this correspondence preserves the von Neumann algebra generated by the representations and also the equivalence of representations</u>.

Since the von Neumann algebra generated by a factor representation of a compact group is finite-dimensional , <u>any factor representation of</u> $L(G_\infty)$ <u>which generates an infinite-dimensional</u>

von Neumann algebra (types I_∞, II , III) corresponds automatically to a factor representation of the direct limit group G_∞ .

§ 2 The AF - algebra associated to a direct limit of compact groups and its diagonalisation

Since we are concerned with * - representations of the involutive Banach algebra $L(G_\infty)$, it is natural to consider the envelopping C^*- algebra $A = A(G_\infty)$ of $L(G_\infty)$. As we shall see , $A(G_\infty)$ is an AF - algebra and the aim of this section will be to carry out for this particular AF - algebra some of the constructions outlined in Chapter I .

II.2.1. In general it is a difficult task to determine the envelopping C^*- algebra ([7]) $\tilde{X}$ of a given involutive Banach algebra X . For L , the determination of A becomes much easier by using the following simple remark .

LEMMA . Let X be an involutive Banach algebra and let $\{X_n\}$ be an increasing sequence of finite dimensional involutive subalgebras such that $\overline{\bigcup_{n=1}^{\infty} X_n} = X$. Suppose moreover that for each $n \in \mathbb{N}$ there is a * - representation π_n of X whose restriction to X_n is faithful . Then $\tilde{X}$ is the direct limit of the $\tilde{X}_n$ ' s .

The proof is obvious , so we omit it .

II.2.2. For X = L we shall construct a special sequence

First some notations. In order to avoid notational complications in the sequel we shall denote the convolution as a usual multiplication :

$$a * b = a\, b \quad .$$

Let $\hat{G}_n$ be the dual of the separable compact group G_n , i.e. the set of equivalence classes of irreducible unitary representations of G_n . For $\rho_n \in \hat{G}_n$ we denote by

$$\chi_{\rho_n} : G_n \longrightarrow \mathbb{C}$$

its character , by

$$d_{\rho_n} = \chi_{\rho_n}(e)$$

its dimension and by $\bar{\rho}_n$ the corresponding conjugate representation . Then

$$p_{\rho_n} = d_{\rho_n} \chi_{\bar{\rho}_n} \in L^1(G_n)$$

is a central projection (i.e. selfadjoint idempotent) in

$$M(G_n) \supset L_{(n)} \quad .$$

We write

$$\rho_n \prec \rho_m \qquad (\rho_n \in \hat{G}_n , \rho_m \in \hat{G}_m , n < m)$$

if ρ_n appears in the restriction of ρ_m to G_n . Then we have

$$\rho_n \prec \rho_m \Longleftrightarrow p_{\rho_n} p_{\rho_m} \neq 0$$

and

$$p_{\rho_m} = \left(\sum_{\rho_n \in \hat{G}_n , \rho_n \prec \rho_m} p_{\rho_n} \right) p_{\rho_m} = \sum_{\rho_n \in \hat{G}_n} p_{\rho_n} p_{\rho_m} \quad .$$

We denote by

$$[\rho_{n+1} : \rho_n]$$

the multiplicity of ρ_n in the restriction of ρ_{n+1} to G_n .

We can find a sequence $\{\Delta_n\}_{n=0,1,2,\ldots}$ of finite sets

$$\Delta_n \subset \bigcup_{k=0}^{n} \hat{G}_k$$

enjoing the following properties

(i) $\Delta_n \subset \Delta_{n+1}$

(ii) $\bigcup_{n=0}^{\infty} \Delta_n = \bigcup_{k=0}^{\infty} \hat{G}_k$

(iii) $\rho_j \prec \rho_{j+1}\ ,\ \rho_{j+1} \in \Delta_n \Longrightarrow \rho_j \in \Delta_n$.

Then we can take

$$X_n = \sum_{\rho_j \in \Delta_n} B_{\rho_j} \quad ,$$

where

$$B_{\rho_j} = p_{\rho_j} L^1(G_j) = p_{\rho_j} M(G_j) = p_{\rho_j} L_{(j)} \quad .$$

The B_{ρ_j} ' s are finite-dimensional involutive subalgebras , so X_n is finite-dimensional and selfadjoint . Moreover , for $j \leqslant k$ we have

$$B_{\rho_j} B_{\rho_k} \subset M(G_j)\, p_{\rho_k} M(G_k) \subset p_{\rho_k} M(G_k) = B_{\rho_k} \quad ,$$

so X_n is also a subalgebra . By the Peter-Weyl theorem

$$\sum_{\rho_j \in \hat{G}_j} B_{\rho_j} \quad \text{is dense in } L^1(G_j) \quad ,$$

hence

$$\bigcup_{n=0}^{\infty} X_n = \sum_{j=0}^{\infty} \sum_{\rho_j \in \hat{G}_j} B_{\rho_j}$$

is dense in $X = L$. Consider λ_n the regular representation of G_n on $L^2(G_n)$. Then the extension of λ_n to $M(G_n) \supset X_n$ is faithful and so , taking $\pi_n = \pi_{\lambda_n}$ (the representation of L corresponding to λ_n by the construction in II.1.2.) , the last condition of Lemma II.2.1 is also satisfied.

Therefore the envelopping C^*- algebra $A = \widetilde{L}$ is the direct limit of the C^*- algebras $A_n = \widetilde{X}_n$. Since X_n is finite-dimensional , A_n is equal to X_n as involutive algebras and we have

$$A = \left\langle \bigcup_{n=0}^{\infty} A_n \right\rangle \quad .$$

In particular , A is an AF - algebra.

For any element $x \in L$ we shall denote by the same symbol its canonical image in $\widetilde{L} = A$.

II.2.3. For further investigations it is necessary to decompose the algebras X_n into factors .

We define the selfadjoint idempotents

$$p_j^{(n)} = \sum_{\rho_j \in \Delta_n \cap \hat{G}_j} p_{\rho_j} \quad \in \; L^1(G_j) \cap X_n$$

and

$$q_{\rho_j}^{(n)} = p_{\rho_j}(1 - p_{j+1}^{(n)}) \quad \in \quad M(G_{j+1}) \cap X_n \quad ; \quad \rho_j \in \Delta_n \quad .$$

The condition (iii) satisfied by the Δ_n ' s implies

$$p_j^{(n)} \geqslant p_{j+1}^{(n)} \quad ,$$

while condition (i) implies

$$p_j^{(n)} \leqslant p_j^{(n+1)} \quad .$$

Since $p_{\rho_j} \in L^1(G_j)$, $p_{\rho_j} p_{j+1}^{(n)} \in L^1(G_{j+1})$ and $L^1(G_j) \cap L^1(G_{j+1}) = \{0\}$, it follows that

$$q_{\rho_j}^{(n)} \neq 0 \quad \text{for each } \rho_j \in \Delta_n \quad .$$

LEMMA. <u>The decomposition of</u> X_n <u>into factors is</u>

$$X_n = \sum_{\rho_j \in \Delta_n} q^{(n)}_{\rho_j} B_{\rho_j}$$

<u>Proof</u>. If $j < k$ we have

$$q^{(n)}_{\rho_j} p_{\rho_k} = p_{\rho_j} (1 - p^{(n)}_{j+1}) p_{\rho_k}$$

$$= p_{\rho_j} (1 - p^{(n)}_{j+1}) \left(\sum_{\rho_{j+1} \in \hat{G}_{j+1} \cap \Delta_n} p_{\rho_{j+1}} p_{\rho_k} \right) = 0 .$$

Therefore

$$j < k \Longrightarrow q^{(n)}_{\rho_j} p_{\rho_k} = 0 = p_{\rho_k} q^{(n)}_{\rho_j}$$

and in particular

$$j \neq k \Longrightarrow q^{(n)}_{\rho_j} q^{(n)}_{\rho_k} = 0 .$$

It is obvious that

$$\rho_j \neq \rho'_j \Longrightarrow q^{(n)}_{\rho_j} q^{(n)}_{\rho'_j} = 0 .$$

Hence $\left\{ q^{(n)}_{\rho_j} \right\}_{\rho_j \in \Delta_n}$ are mutually orthogonal. Then we have

$$\sum_{\rho_j \in \Delta_n} q^{(n)}_{\rho_j} = \sum_{j=0}^{n} \sum_{\rho_j \in \Delta_n \cap \hat{G}_j} p_{\rho_j} (1 - p^{(n)}_{j+1})$$

$$= \sum_{j=0}^{n} p^{(n)}_j (1 - p^{(n)}_{j+1})$$

$$= \sum_{j=0}^{n} \left(p^{(n)}_j - p^{(n)}_{j+1} \right)$$

$$= p^{(n)}_0 - p^{(n)}_{n+1}$$

$$= 1 .$$

Moreover, by their definition, $q^{(n)}_{\rho_j}$ are central elements in $M(G_j)$ hence for $x \in B_{\rho_k}$, $k \leqslant j$, we have

$$q^{(n)}_{\rho_j} x = x q^{(n)}_{\rho_j} .$$

On the other hand for $x \in B_{\rho_k}$, $k > j$, we have

$$q^{(n)}_{\rho_j} x = q^{(n)}_{\rho_j} p_{\rho_k} x = 0 ,$$

$$x q^{(n)}_{\rho_j} = x p_{\rho_k} q^{(n)}_{\rho_j} = 0 .$$

Hence $\left\{q^{(n)}_{\rho_j}\right\}_{\rho_j \in \Delta_n}$ are central elements in X_n. Finally, the map

$$B_{\rho_j} \ni x \longmapsto q^{(n)}_{\rho_j} x \in q^{(n)}_{\rho_j} B_{\rho_j} = q^{(n)}_{\rho_j} X_n$$

is a non-zero involutive homomorphism of the factor B_{ρ_j}, so that $q^{(n)}_{\rho_j} B_{\rho_j}$ is itself a factor.

Q.E.D.

Consequently, the decomposition into factors of the finite-dimensional C^*- algebra A_n is

$$A_n = \sum_{\rho_j \in \Delta_n} q^{(n)}_{\rho_j} B_{\rho_j} ,$$

i.e. $\left\{q^{(n)}_{\rho_j}\right\}_{\rho_j \in \Delta_n}$ are the minimal central projections in A_n.

II.2.4. Now we construct a maximal system of minimal projections in A_n and a corresponding system of matrix units for A_n.

The idea behind our construction is that of the Gelfand - Zeitlin basis, i.e. first we write the representation $\rho_{n+1} \mid G_n$

as a sum of irreducible representations of G_n , then we split again these representations into irreducible representations of G_{n-1} and so on . Since $G_0 = \{e\}$, at the end we find one-dimensional spaces . However , this will be done algebraically in A_n .

Fix the representations $\rho_n \in \hat{G}_n$, $\rho_{n+1} \in \hat{G}_{n+1}$, $\rho_n \prec \rho_{n+1}$. Then

$$p_{\rho_n} B_{\rho_{n+1}} p_{\rho_n} = (p_{\rho_n} p_{\rho_{n+1}}) B_{\rho_{n+1}} (p_{\rho_n} p_{\rho_{n+1}}) \subset B_{\rho_{n+1}}$$

is a factor of type $I_{(d_{\rho_n}[\rho_{n+1}:\rho_n])}$. Since the map

$$B_{\rho_n} \ni x \longmapsto p_{\rho_{n+1}} x \in p_{\rho_{n+1}} B_{\rho_n}$$

is a non-zero involutive homomorphism , it follows that $p_{\rho_{n+1}} B_{\rho_n}$ is a factor of type $I_{d_{\rho_n}}$. Moreover ,

$$p_{\rho_{n+1}} B_{\rho_n} = p_{\rho_{n+1}} p_{\rho_n} B_{\rho_n} p_{\rho_n} p_{\rho_{n+1}} \subset p_{\rho_n} B_{\rho_{n+1}} p_{\rho_n} \quad .$$

Hence the commutant of $p_{\rho_{n+1}} B_{\rho_n}$ in $p_{\rho_n} B_{\rho_{n+1}} p_{\rho_n}$ is a factor of type $I_{[\rho_{n+1}:\rho_n]}$. We choose a maximal system

$$p_{(\rho_n \xrightarrow{k} \rho_{n+1})} \quad , \quad 1 \leqslant k \leqslant [\rho_{n+1}:\rho_n] \quad ,$$

of mutually orthogonal minimal projections in

$$(p_{\rho_{n+1}} B_{\rho_n})' \cap (p_{\rho_n} B_{\rho_{n+1}} p_{\rho_n}) \quad .$$

Of course ,

$$\sum_{k=1}^{[\rho_{n+1}:\rho_n]} p_{(\rho_n \xrightarrow{k} \rho_{n+1})} = p_{\rho_n} p_{\rho_{n+1}} \quad .$$

We denote by $S(\rho_n)$, $\rho_n \in \hat{G}_n$, the set of all symbols

$$\rho_0 \xrightarrow{k_1} \rho_1 \longrightarrow \cdots \longrightarrow \rho_{j-1} \xrightarrow{k_j} \rho_j \longrightarrow \cdots \longrightarrow \rho_{n-1} \xrightarrow{k_n} \rho_n$$

where

$$\rho_j \in \hat{G}_j \quad , \quad \rho_j \prec \rho_{j+1} \quad , \quad 1 \leq k_j \leq [\rho_j : \rho_{j-1}]$$

and we define

$$p\left(\rho_0 \xrightarrow{k_1} \cdots \xrightarrow{k_n} \rho_n\right) = \prod_{j=1}^{n} p\left(\rho_{j-1} \xrightarrow{k_j} \rho_j\right)$$

Then

$$\{p_\alpha\}_{\alpha \in S(\rho_n)}$$

is a maximal system of mutually orthogonal minimal projections in B_{ρ_n} .

Indeed , this is obvious for $n = 0$. Suppose this is true for all $\rho_n \in \hat{G}_n$, with $n \in \mathbb{N}$ fixed , and consider $\rho_{n+1} \in \hat{G}_{n+1}$. Then

$$\left\{p_\alpha \, p_{\rho_{n+1}}\right\}_{\alpha \in S(\rho_n)}$$

is a maximal set of minimal projections in $p_{\rho_{n+1}} B_{\rho_n}$. Since

$$\left\{p\left(\rho_n \xrightarrow{k} \rho_{n+1}\right)\right\}_{1 \leq k \leq [\rho_{n+1} : \rho_n]}$$

is a maximal set of minimal projections in the commutant of $p_{\rho_{n+1}} B_{\rho_n}$ in $p_{\rho_n} B_{\rho_{n+1}} p_{\rho_n}$, it follows that

$$\left\{p_\alpha \, p\left(\rho_n \xrightarrow{k} \rho_{n+1}\right)\right\}_{\alpha \in S(\rho_n) \, , \, 1 \leq k \leq [\rho_{n+1} : \rho_n]}$$

is a maximal set of minimal projections in $p_{\rho_n} B_{\rho_{n+1}} p_{\rho_n}$. Hence

$$\{p_\beta\}_{\beta \in S(\rho_{n+1})}$$

is a maximal set of minimal projections in $B_{\rho_{n+1}}$.

For $\alpha \in S(\rho_n)$ and $1 \leq k \leq [\rho_{n+1}:\rho_n]$ we consider the symbols $\alpha \xrightarrow{k} \rho_{n+1}$ in $S(\rho_{n+1})$. We have

$$\sum_{\alpha \in S(\rho_n)\ ,\ 1 \leq k \leq [\rho_{n+1}:\rho_n]} p_{(\alpha \xrightarrow{k} \rho_{n+1})} = p_{\rho_n} p_{\rho_{n+1}} .$$

II.2.5. We choose now by induction a system of matrix units

$$\{E_{\alpha,\beta}\}_{\alpha,\beta \in S(\rho_n)}$$

in B_{ρ_n} . Suppose this was done for all $\rho_n \in \hat{G}_n$, with $n \in \mathbb{N}$ fixed, and consider $\rho_{n+1} \in \hat{G}_{n+1}$. Then

$$\left\{p_{(\rho_n \xrightarrow{k} \rho_{n+1})} E_{\alpha,\beta}\, p_{(\rho_n \xrightarrow{k} \rho_{n+1})}\right\}_{\alpha,\beta \in S(\rho_n)}$$

is a system of matrix units in

$$p_{(\rho_n \xrightarrow{k} \rho_{n+1})} B_{\rho_n}\, p_{(\rho_n \xrightarrow{k} \rho_{n+1})}$$

so that

$$\left\{p_{(\rho_n \xrightarrow{k} \rho_{n+1})} E_{\alpha,\beta}\, p_{(\rho_n \xrightarrow{k} \rho_{n+1})}\right\}_{\alpha,\beta \in S(\rho_n)\ ,\ 1 \leq k \leq [\rho_{n+1}:\rho_n]\ ;\ \rho_n \prec \rho_{n+1}}$$

is a system of matrix units in the involutive subalgebra

$$\sum_{\rho_n \prec \rho_{n+1},\ 1 \leq k \leq [\rho_{n+1}:\rho_n]} p_{(\rho_n \xrightarrow{k} \rho_{n+1})} B_{\rho_n}\, p_{(\rho_n \xrightarrow{k} \rho_{n+1})}$$

of $B_{\rho_{n+1}}$. We choose an arbitrary completion of this system to a system of matrix units in $B_{\rho_{n+1}}$.

The systems of matrix units constructed as above have the property that each

$$p_{\rho_{n+1}} E_{\alpha,\beta} \quad ; \quad \alpha, \beta \in S(\rho_n)$$

is expressed with respect to the linear base $\{E_{\mu,\nu}\}_{\mu,\nu \in S(\rho_{n+1})}$ of $B_{\rho_{n+1}}$ as a linear combination with coefficients only 0 and 1. Namely

$$(1) \qquad p_{\rho_{n+1}} E_{\alpha,\beta} = \sum_{k=1}^{[\rho_{n+1}:\rho_n]} E_{(\alpha \xrightarrow{k} \rho_{n+1}),(\beta \xrightarrow{k} \rho_{n+1})} .$$

II.2.6. Denote by Perm $S(\rho_n)$ the set of all the permutations of $S(\rho_n)$. For any $\sigma \in \text{Perm } S(\rho_n)$ we define

$$V_\sigma = \sum_{\alpha \in S(\rho_n)} E_{\alpha,\sigma(\alpha)} \in B_{\rho_n} ,$$

$$U_\sigma = V_\sigma + (1 - p_{\rho_n}) \in X_n .$$

Let us fix $\rho_n \prec \rho_{n+1}$ and $\sigma \in \text{Perm } S(\rho_n)$. Since any symbol in $S(\rho_{n+1})$ is of the form

$$\alpha \xrightarrow{k} \rho_{n+1} \quad \text{with} \quad \alpha \in S(\rho_n') , 1 \leq k \leq [\rho_{n+1}:\rho_n'] , \rho_n' \prec \rho_{n+1}$$

we can define $\tau \in \text{Perm } S(\rho_{n+1})$ by

$$\tau(\alpha \xrightarrow{k} \rho_{n+1}) = \alpha \xrightarrow{k} \rho_{n+1} \quad \text{if } \alpha \in S(\rho_n') , \rho_n' \neq \rho_n ,$$

$$\tau(\alpha \xrightarrow{k} \rho_{n+1}) = \sigma(\alpha) \xrightarrow{k} \rho_{n+1} \quad \text{if } \alpha \in S(\rho_n) .$$

Then we have

(2) $$p_{\rho_{n+1}} U_{\sigma} = p_{\rho_{n+1}} U_{\tau} .$$

Indeed ,

$$p_{\rho_{n+1}} U_{\tau} = p_{\rho_{n+1}} \sum_{\beta \in S(\rho_{n+1})} E_{\beta,\tau(\beta)} =$$

$$= p_{\rho_{n+1}} \sum_{\alpha \in S(\rho_n),\ 1 \leqslant k \leqslant [\rho_{n+1}:\rho_n]} E_{(\alpha \xrightarrow{k} \rho_{n+1}),(\sigma(\alpha) \xrightarrow{k} \rho_{n+1})} +$$

$$+ p_{\rho_{n+1}} \sum_{\substack{\alpha \in S(\rho_n'),\ 1 \leqslant k \leqslant [\rho_{n+1}:\rho_n] \\ \rho_n' \prec \rho_{n+1},\ \rho_n' \neq \rho_n}} E_{(\alpha \xrightarrow{k} \rho_{n+1}),(\alpha \xrightarrow{k} \rho_{n+1})}$$

$$= p_{\rho_{n+1}} \sum_{\alpha \in S(\rho_n)} E_{\alpha,\sigma(\alpha)} + p_{\rho_{n+1}} \sum_{\alpha \in S(\rho_n'),\rho_n' \prec \rho_{n+1},\rho_n' \neq \rho_n} E_{\alpha,\alpha}$$

$$= p_{\rho_{n+1}} V_{\sigma} + p_{\rho_{n+1}} \sum_{\rho_n' \prec \rho_{n+1},\ \rho_n' \neq \rho_n} p_{\rho_n'} =$$

$$= p_{\rho_{n+1}} V_{\sigma} + p_{\rho_{n+1}} (1 - p_{\rho_n}) =$$

$$= p_{\rho_{n+1}} U_{\sigma} .$$

II.2.7. Concerning the AF - algebra A we first remark that

$$\left\{ p_{\alpha}\, q_{\rho_j}^{(n)} \right\}_{\alpha \in S(\rho_j),\ \rho_j \in \Delta_n}$$

is a maximal system of minimal projections in

$$A_n = \sum_{\rho_j \in \Delta_n} q^{(n)}_{\rho_j} B_{\rho_j}$$

and we denote by C_n the corresponding m.a.s.a. in A_n .

Since for every $\alpha \in S(\rho_j)$ we have

$$\begin{aligned} p_\alpha q^{(n)}_{\rho_j} &= p_\alpha p_{\rho_j} (1 - p^{(n)}_{j+1}) \\ &= p_\alpha (1 - \sum_{\rho_{j+1} \in \Delta_n \cap \hat{G}_{j+1}} p_{\rho_{j+1}}) \end{aligned}$$

and since obviously

$$p_{\rho_{j+1}} = \sum_{\beta \in S(\rho_{j+1})} p_\beta ,$$

it follows that C_n is generated by

$$\{p_\alpha\}_{\alpha \in S(\rho_j)\ ,\ \rho_j \in \Delta_n} .$$

Let us denote by D^o_{n+1} the abelian subalgebra of A_{n+1} generated by

$$\left\{p_{(\rho_j \xrightarrow{k} \rho_{j+1})}\right\}_{\rho_j < \rho_{j+1} \in \Delta_{n+1}\ ,\ \rho_{j+1} \notin \Delta_n\ ,\ 1 \leq k \leq [\rho_{j+1} : \rho_j]}$$

It is clear that

$$C_{n+1} = \langle C_n , D^o_{n+1} \rangle .$$

We assert that

$$D^o_{n+1} \subset A'_n \cap A_{n+1} .$$

Indeed , consider $\rho_i \in \Delta_n$, $x \in B_{\rho_i}$ and $\rho_j < \rho_{j+1} \in \Delta_{n+1}$, $\rho_{j+1} \notin \Delta_n$, $1 \leq k \leq [\rho_{j+1} : \rho_j]$. Because of the condition (iii) satisfied by the Δ_n ' s , we cannot have $\rho_{j+1} \prec \rho_i$. Therefore , if $j+1 \leq i$ we have $p_{\rho_{j+1}} p_{\rho_i} = 0$ and so ,

$$p_{(\rho_j \xrightarrow{k} \rho_{j+1})} x = x\, p_{(\rho_j \xrightarrow{k} \rho_{j+1})} = 0 \ .$$

On the other hand, if $i < j+1$ we have

$$x\, p_{\rho_j} p_{\rho_{j+1}} = p_{\rho_j} p_{\rho_{j+1}} x \in p_{\rho_{j+1}} B_{\rho_j} \ , \quad p_{(\rho_j \xrightarrow{k} \rho_{j+1})} \in (p_{\rho_{j+1}} B_{\rho_j})'$$

and so

$$\begin{aligned} x\, p_{(\rho_j \xrightarrow{k} \rho_{j+1})} &= x\, p_{\rho_j} p_{\rho_{j+1}} p_{(\rho_j \xrightarrow{k} \rho_{j+1})} \\ &= p_{(\rho_j \xrightarrow{k} \rho_{j+1})} p_{\rho_j} p_{\rho_{j+1}} x \\ &= p_{(\rho_j \xrightarrow{k} \rho_{j+1})} x \ . \end{aligned}$$

Putting

$$D_{n+1} = C_{n+1} \cap A_n' \cap A_{n+1} \ ,$$

it follows that D_{n+1} is a m.a.s.a. in $A_n' \cap A_{n+1}$ and

$$C_{n+1} = \langle C_n \ , \ D_{n+1} \rangle \ .$$

Hence we are in the situation treated in Chapter I , § 1 . In particular,

$$C = \left\langle \bigcup_{n=0}^{\infty} C_n \right\rangle$$

is a m.a.s.a. in A and we have the corresponding conditional expectation

$$P : A \longrightarrow C$$

of A with respect to C .

II.2.8. Denote by Ω_n , Ω the spectra of C_n , C respectively . Of course, Ω_n can be identified with the discrete set of all the minimal projections in C_n , hence with the discrete set of all the symbols

$$\alpha = (\rho_0 \xrightarrow{k_1} \rho_1 \longrightarrow \dots \longrightarrow \rho_{j-1} \xrightarrow{k_j} \rho_j)$$

where $\alpha \in S(\rho_j)$, $\rho_j \in \Delta_n$.

The map $\Omega_{n+1} \longrightarrow \Omega_n$ canonically corresponding to the inclusion $C_n \subset C_{n+1}$ associates to every minimal projection in C_{n+1} the unique minimal projection in C_n containing it . This means that the above map associates to the symbol

$$\rho_0 \xrightarrow{k_1} \rho_1 \longrightarrow \dots \longrightarrow \rho_{j-1} \xrightarrow{k_j} \rho_j \in \Omega_{n+1}$$

the symbol

$$\rho_0 \xrightarrow{k_1} \rho_1 \longrightarrow \dots \longrightarrow \rho_{i-1} \xrightarrow{k_i} \rho_i \in \Omega_n$$

where $i \leq j$ is maximal with the property

$$\rho_i \in \Delta_n \quad .$$

The spectrum Ω of C is the topological inverse limit of the spaces Ω_n and of the maps $\Omega_{n+1} \longrightarrow \Omega_n$. Owing to condition (ii) satisfied by the Δ_n 's , it is easy to see that <u>the set Ω consists of all the symbols</u>

$$t = (\rho_0(t) \xrightarrow{k_1(t)} \rho_1(t) \longrightarrow \dots \longrightarrow \rho_{n-1}(t) \xrightarrow{k_n(t)} \rho_n(t) \longrightarrow \dots)$$

<u>where</u>

$$1 \leq n < n_0(t) \in \mathbb{N} \cup \{+\infty\} \quad ,$$

$$\rho_n(t) \in \hat{G}_n \quad ,$$

$$1 \leqslant k_n(t) \leqslant [\rho_n(t) : \rho_{n-1}(t)] ;$$

thus, $n_0(t) \in \mathbb{N} \cup \{+\infty\}$ indicates how many groups G_n are involved in the symbol of $t \in \Omega$.

Bearing in mind the very definition of the inverse limit topology and the special nature of the maps $\Omega_{n+1} \longrightarrow \Omega_n$, one finds the following <u>description of the topology on</u> Ω .

Let be $\omega \subset \Omega$ and $t \in \Omega$. Then

$$t \in \overline{\omega}$$

<u>if and only if any of the following conditions holds</u> :

(i) $t \in \omega$;

(ii) $n_0(t) = +\infty$ <u>and for every</u> $m \in \mathbb{N}$ <u>there is</u> $s \in \omega$ <u>such that</u>

$$\rho_n(s) = \rho_n(t) \quad , \quad k_n(s) = k_n(t)$$

<u>for any</u> $n \leqslant m$;

(iii) $n_0(t) < +\infty$ <u>and the set</u>

$$\left\{ \rho_{n_0(t)}(s) \quad ; \quad s \in \omega ,\ \rho_n(s) = \rho_n(t) ,\ k_n(s) = k_n(t) ,\ (\forall)\ n < n_0(t) \right\}$$

<u>is infinite</u>.

II.2.9. Now we remark that, for each $\rho_j \in \Delta_n$,

$$\left\{ q^{(n)}_{\rho_j} E_{\alpha,\beta} \right\}_{\alpha,\beta \in S(\rho_j)}$$

is a system of matrix units in $q^{(n)}_{\rho_j} B_{\rho_j}$, hence

$$\left\{ q^{(n)}_{\rho_j} E_{\alpha,\beta} \right\}_{\alpha,\beta \in S(\rho_j) \ ; \ \rho_j \in \Delta_n}$$

is a system of matrix units in A_n . Using the relation (1) it is easy to verify that each element of the system of matrix units in A_n is written as a linear combination with coefficients only 0 and 1 in the linear base of A_{n+1} given by its considered system of matrix units.

As in the case of general AF - algebras we denote by U_n the group of unitary elements in A_n associated with the above system of matrix units in A_n and we have

$$U_n \subset U_{n+1} \qquad ,$$

which allows us to consider the group

$$U = \bigcup_{n=0}^{\infty} U_n \subset A \qquad .$$

The group U_n is generated by the elements

$$q^{(n)}_{\rho_j} U_{\sigma} + (1 - q^{(n)}_{\rho_j}) \quad ; \quad \sigma \in \text{Perm } S(\rho_j) \quad , \quad \rho_j \in \Delta_n \qquad .$$

Fix such an unitary in U_n and denote by

$$\rho^{i}_{j+1} \quad , \quad i = 1,\dots,m$$

the elements of the set $\left\{ \rho_{j+1} \in \Delta_n \cap \hat{G}_{j+1} \quad ; \quad \rho_j \prec \rho_{j+1} \right\}$. By using (2) we find permutations

$$\tau_i \in \text{Perm } S(\rho^{i}_{j+1}) \quad , \quad i = 1,\dots,m$$

such that

$$p_{\rho^{i}_{j+1}} U_{\sigma} = p_{\rho^{i}_{j+1}} U_{\tau_i} \ , \quad i = 1,\dots,m \qquad .$$

It follows that

$$p^{(n)}_{j+1} U_{\sigma} = \sum_{i=1}^{m} p_{\rho^{i}_{j+1}} U_{\sigma}$$

$$= \sum_{i=1}^{m} p_{\rho^{i}_{j+1}} U_{\tau_i}$$

$$= \sum_{i=1}^{m} p_{\rho^{i}_{j+1}} U_{\tau_1} \cdots U_{\tau_m}$$

$$= p_{j+1}^{(n)} \; U_{\tau_1} \dots U_{\tau_m}$$

Put $W = U_{\tau_1} \dots U_{\tau_m}$. Then we have

$$p_{j+1}^{(n)} U_\sigma = p_{j+1}^{(n)} W ,$$

$$U_\sigma = p_{\rho_j} U_\sigma + (1 - p_{\rho_j}) \quad , \quad W = p_{j+1}^{(n)} W + (1 - p_{j+1}^{(n)})$$

and also similar formulas with U_σ^{-1} , W^{-1} . Therefore

$$(1 - p_{\rho_j}) \, p_{j+1}^{(n)} W^{-1} = (1 - p_{\rho_j}) \, p_{j+1}^{(n)} U^{-1} = (1 - p_{\rho_j}) \, p_{j+1}^{(n)}$$

and so

$$\begin{aligned} U_\sigma W^{-1} &= (p_{\rho_j} U + (1 - p_{\rho_j}))(p_{j+1}^{(n)} W^{-1} + (1 - p_{j+1}^{(n)})) \\ &= p_{\rho_j} p_{j+1}^{(n)} + (1 - p_{\rho_j}) \, p_{j+1}^{(n)} + p_{\rho_j} (1 - p_{j+1}^{(n)}) \, U_\sigma + \\ &\qquad + 1 - p_{j+1}^{(n)} - p_{\rho_j} (1 - p_{j+1}^{(n)}) \\ &= q_{\rho_j}^{(n)} U_\sigma + (1 - q_{\rho_j}^{(n)}) \end{aligned}$$

The above discusion shows that <u>the group</u> U <u>is generated by the elements</u>

$$U_\sigma \; ; \; \sigma \in \text{Perm } S(\rho_n) \quad , \quad \rho_n \in \hat{G}_n \quad , \quad n \in \mathbb{N} .$$

We know from the general case of an AF - algebra treated in Chapter I § 1 that the group U invariantes the m.a.s.a. C , i.e.

$$u \, C \, u^* = C \qquad \text{for every} \qquad u \in U$$

and that U induces a group Γ of homeomorphisms of Ω.

From the preceding remarks we infer the following <u>concrete description of</u> Γ . Fix $n \in \mathbb{N}$, $\rho_n \in \hat{G}_n$ and $\sigma \in \text{Perm } S(\rho_n)$ and <u>define a transformation</u>

$$\gamma = \gamma(n, \rho_n, \sigma)$$

<u>of</u> Ω <u>as follows</u> :

$$\gamma(t) = \left(\sigma\left(\rho_0(t) \xrightarrow{k_1(t)} \dots \xrightarrow{k_n(t)} \rho_n(t)\right) \xrightarrow{k_{n+1}(t)} \rho_{n+1}(t) \dots\right)$$

<u>if</u> $t \in \Omega$, $n_0(t) > n$ <u>and</u> $\rho_n(t) = \rho_n$, <u>and</u>

$$\gamma(t) = t$$

<u>in the contrary case. Then</u> Γ <u>is the group generated by the transformations</u> :

$$\gamma(n, \rho_n, \sigma) \quad , \sigma \in \text{Perm } S(\rho_n) \ , \ \rho_n \in \hat{G}_n \quad , \quad n \in \mathbb{N} \ .$$

Thus, we have fulfiled our task of describing suitable $\Omega = \Omega(G_\infty)$ and $\Gamma = \Gamma(G_\infty)$ for the AF - algebra $A(G_\infty)$ of a direct limit G_∞ of compact groups G_n . In concrete situations, all we need to know in order to construct $\Omega(G_\infty)$ and $\Gamma(G_\infty)$ are the sets $\hat{G}_n$ and the numbers $[\rho_{n+1} : \rho_n]$.

Let us recall that the isomorphism between A and $A(\Omega, \Gamma)$ established in Chapter I § 1 is not unique , it does depend on the choice of the $E_{\alpha,\beta}$'s (even in the case when $[\rho_{n+1} : \rho_n] \leq 1$ for all ρ_n , ρ_{n+1}) .

II.2.10. For the applications we have in mind we must explicitate when the representation of A corresponding to a given representation of G_∞ is a representation π_μ corresponding , as in Chapter I § 3 , to a measure μ on Ω . This will be done by using the special state $\mu \circ P$ associated to π_μ .

Thus , let ρ be a representation of G_∞ on a Hilbert space H and $\xi \in H$, $\|\xi\| = 1$, a cyclic vector for ρ . Denote by π the representation π_ρ of L associated to ρ as in Section II.1.2. and use the same symbol for the corresponding representation of $A = \tilde{L}$. Then $\pi \mid L^1(G_n)$ is the representation of $L^1(G_n)$ corresponding to $\rho \mid G_n$. Consider a Γ- quasi-invariant measure μ on Ω.

In order that π be equivalent to π_μ and the state of A associated to π and ξ be equal to $\mu \circ P$, it is necessary and sufficient that

$$(3) \qquad (\pi(x)\xi \mid \xi) = \mu(P(x))$$

for all $x \in L^1(G_n)$ and $n \in \mathbb{N}$. Since the B_{ρ_n}'s span $L^1(G_n)$, and each B_{ρ_n} is spanned by the elements of the form

$$x = \delta_g * p_{\rho_n} \quad , \qquad g \in G_n \quad ,$$

it is sufficient to verify (3) only for these x.

On the other hand, we have

$$\mu(P(x)) = \sum_{\alpha \in S(\rho_n)} \mu(p_\alpha x \, p_\alpha)$$

$$= \sum_{\alpha \in S(\rho_n)} \mu(p_\alpha) \, \mathrm{Tr}_{B_{\rho_n}}(p_\alpha x \, p_\alpha) \quad .$$

Therefore, <u>in order that</u> $\pi \simeq \pi_\mu$ <u>and</u> $\omega_\xi \circ \pi = \mu \circ P$, <u>it is necessary and sufficient that</u>

$$(\pi(\delta_g * p_{\rho_n})\xi \mid \xi) = \sum_{\alpha \in S(\rho_n)} \mu(p_\alpha) \, \mathrm{Tr}_{B_{\rho_n}}(p_\alpha * \delta_g * p_\alpha)$$

<u>for all</u> $g \in G_n$, $\rho_n \in \hat{G}_n$ <u>and</u> $n \in \mathbb{N}$.

CHAPTER III THE PRIMITIVE IDEALS OF $A(U(\infty))$

We consider now a special case of the above general situation , namely the group $U(\infty)$ which caused the preceding discussion . While the irreducible representations of the unitary groups $U(n)$ are completely described in terms of signatures by the classical theory of Hermann Weyl , it seems difficult to classify the irreducible representations of $U(\infty)$. In this Chapter we shall study in some detail the primitive ideals of the C^*- algebra $A(U(\infty))$.

§ 1 The primitive spectrum of $A(U(\infty))$

The aim of this section is to give a complete description of the primitive spectrum of the AF - algebra $A = A(U(\infty))$. As we shall see , a primitive ideal of A is determined by an "upper signature" and a "lower signature" .

III.1.1. By the classical theory of Hermann Weyl , the dual $U(n)$ of the unitary group $U(n)$ can be identified with the discrete set of all the signatures

$$m_1 \geqslant m_2 \geqslant \dots \geqslant m_n$$

where m_i Z , $i = 1, 2, \dots, n$.

Then , for

$$\rho_{n-1} = (m_1^{(n-1)} \geqslant m_2^{(n-1)} \geqslant \dots \geqslant m_{n-1}^{(n-1)}) \in \widehat{U(n-1)}$$

$$\rho_n = (m_1^{(n)} \geqslant m_2^{(n)} \geqslant \dots \geqslant m_n^{(n)}) \in \widehat{U(n)}$$

we have

$$\rho_{n-1} \prec \rho_n$$

if and only if

$$m_{j-1}^{(n)} \geqslant m_{j-1}^{(n-1)} \geqslant m_j^{(n)} \quad , \quad 1 \leqslant j \leqslant n \quad ,$$

and , in this case ,

$$[\rho_n : \rho_{n-1}] = 1 \quad .$$

Taking into account these classical facts and the results of Section II.2.8. , it follows that the points of the space

$$\Omega = \Omega(U(\infty))$$

are the symbols

$$t = \left\{ (m_j^{(n)}(t))_{1 \leqslant j \leqslant n} \right\}_{1 \leqslant n < n_0(t)}$$

where

$$1 \leqslant n_0(t) \leqslant +\infty \quad ,$$

$$m_j^{(n)}(t) \in \mathbb{Z} \quad ,$$

$$m_{j-1}^{(n)}(t) \geqslant m_{j-1}^{(n-1)}(t) \geqslant m_j^{(n)}(t) \quad .$$

Therefore , a point of Ω looks like the picture below , where $a \longrightarrow b$ means $a \leqslant b$.

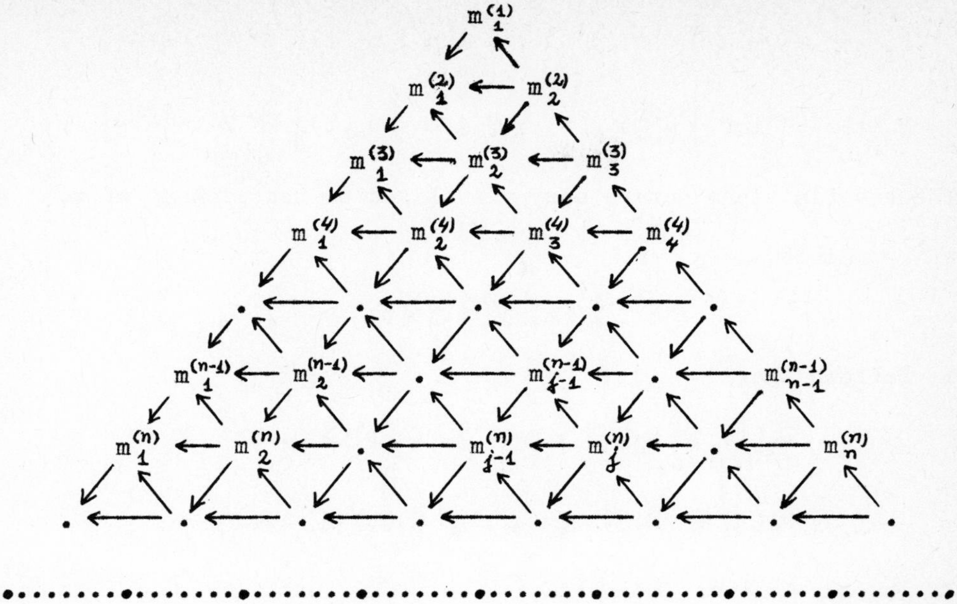

The description of the topology on Ω and the description of the transformation group $\Gamma = \Gamma(U(\infty))$ follow obviously from Sections II.2.8. and II.2.9. Roughly speaking, a point $t \in \Omega$ with $n_0(t) = +\infty$ is adherent to a set $\omega \subset \Omega$ if for every "line" one can find a point in ω with the same "beginning" until that line. Also, the generators of the group Γ change these beginnings among themselves, leaving fixed the rest of the picture.

III.1.2. We recall from Chapter I § 3 that the primitive ideals of A correspond in a canonical way to the closures of the orbits of Γ. In Lemma III.1.3. we shall determine these sets.

First, some notations; for $t \in \Omega$ and for $1 \leq j < n_0(t)$ we define

$$L_j(t) = \sup \left\{ m_j^{(n)}(t) \quad ; j \leqslant n < n_0(t) \right\} \in \mathbb{Z} \cup \{+\infty\} ,$$

$$M_j(t) = \inf \left\{ m_{n-j+1}^{(n)}(t) \; ; j \leqslant n < n_0(t) \right\} \in \mathbb{Z} \cup \{-\infty\} .$$

These definitions can be easy visualised on the picture of t.

Since

$$m_{j-1}^{(n-1)}(t) \geqslant m_j^{(n)}(t) ,$$

it follows that

$$L_1(t) \geqslant L_2(t) \geqslant \dots \geqslant L_j(t) \geqslant \dots$$

$$M_1(t) \leqslant M_2(t) \leqslant \dots \leqslant M_j(t) \leqslant \dots$$

If $n_0(t) < +\infty$, we have

$$+\infty > L_j(t) = m_j^{(n_0(t)-1)}(t) = M_{n_0(t)-j}(t) > -\infty ,$$

while if $n_0(t) = +\infty$, we have

$$L_j(t) \geqslant m_j^{(n)}(t) \geqslant M_{n-j+1}(t) ,$$

that is

$$L_j(t) \geqslant M_k(t) \quad \text{for every } j,k \in \mathbb{N} .$$

III.1.3. LEMMA . <u>Consider</u> $t_0 \in \Omega$, <u>denote by</u> $\omega = \overline{\Gamma(t_0)}$ <u>the closure of the</u> t_0 - <u>orbit of</u> Γ <u>in</u> Ω <u>and put</u>

$$L_j = L_j(t_0) \; , \; M_j = M_j(t_0) \; ; \; 1 \leqslant j < n_0(t_0) .$$

1). <u>If</u> $n_0(t_0) < +\infty$, <u>then</u>

$$\omega = \left\{ t \in \Omega \; ; \; n_0(t) = n_0(t_0) \; , \; m_j^{(n_0(t)-1)}(t) = L_j \; ; \; 1 \leqslant j < n_0(t) \right\}.$$

2). <u>If</u> $n_0(t_0) = +\infty$, <u>then</u>

$$\omega \cap \left\{ t \in \Omega \; ; \; n_0(t) = +\infty \right\} =$$

$$= \left\{ t \in \Omega \; ; \; n_o(t) = +\infty \, , \; L_j \geq m_j^{(n)}(t) \geq M_{n-j+1} \; ; \; 1 \leq j \leq n < +\infty \right\},$$

$$\omega \cap \left\{ t \in \Omega \; ; \; n_o(t) < +\infty \right\} =$$

$$= \begin{cases} \emptyset & \text{if } \; L_1 - M_1 < +\infty \, , \\ \left\{ t \in \Omega \; ; \; n_o(t) < +\infty \, , \; L_j \geq m_j^{(n)}(t) \geq M_{n-j+1} \; ; \; 1 \leq j \leq n < n_o(t) \right\} & \text{if } \; L_1 - M_1 = +\infty \, . \end{cases}$$

<u>Proof</u>. The first statement is obvious. In this case we have

$$\omega = \Gamma(t_o) \; .$$

In order to prove the second statement , denote

$$\Theta = \left\{ t \in \Omega \; ; \; n_o(t) = +\infty \, , \; L_j \geq m_j^{(n)}(t) \geq M_{n-j+1} \; ; \; 1 \leq j \leq n < +\infty \right\}.$$

Obviously ,

$$(1) \qquad \Gamma(t_o) \subset \Theta \; .$$

We shall show that

$$(2) \qquad \Theta \subset \overline{\Gamma(t_o)} = \omega$$

by proving , for any $s \in \Theta$, any $n \in \mathbb{N}$ and any $1 \leq j \leq n$, the following assertion , which we denote by $Af(s;n,j)$:

$$\left[\begin{array}{c} \underline{\text{there exists}} \;\; t \in \Gamma(t_o) \;\; \underline{\text{such that}} \\ m_i^{(k)}(t) = m_i^{(k)}(s) \\ \underline{\text{for every}} \;\; k \;\; \underline{\text{and}} \;\; i \;\; \underline{\text{satisfying}} \\ 1 \leq i \leq k \leq n-1 \;\; \underline{\text{or}} \;\; k = n \, , \; 1 \leq i \leq j \, . \end{array} \right]$$

In order to prove $Af(s;n,j)$ we fix $s \in \Theta$ and we proceed by induction as follows :

(i) $Af(s;1,1)$ is true ,

(ii) $Af(s;n,j) \Longrightarrow Af(s;n,j+1)$, $1 \leq j < n$,

(iii) $Af(s;n,n) \Longrightarrow Af(s;n+1,1)$.

Let us prove (i) . There are two possibilities :

A). $m_1^{(1)}(t_o) \geqslant m_1^{(1)}(s)$ or B). $m_1^{(1)}(t_o) \leqslant m_1^{(1)}(s)$.

First we suppose that

$$A).\ m_1^{(1)}(t_o) \geqslant m_1^{(1)}(s) \quad .$$

Since $s \in \Theta$ we have

$$m_1^{(1)}(s) \geqslant M_1 = \inf \left\{ m_n^{(n)}(t_o)\ ;\ n \geqslant 1 \right\} ,$$

so there exists a unique $h \geqslant 1$ such that

$$m_h^{(h)}(t_o) \geqslant m_1^{(1)}(s) \quad \text{and} \quad m_{h+1}^{(h+1)}(t_o) \leqslant m_1^{(1)}(s) \quad .$$

We define

$$\underline{m}_i^{(k)} = \begin{cases} m_1^{(1)}(s) & \text{if} \quad i = k \leqslant h \quad , \\ m_i^{(k)}(t_o) & \text{if} \quad i \neq k \quad \text{or} \quad k > h \quad . \end{cases}$$

Then $Af(s;1,1)$ holds with

$$t = \left\{ (\underline{m}_i^{(k)})_{1 \leqslant i \leqslant k} \right\}_{1 \leqslant k < +\infty} \in \Omega \quad ,$$

since

$$m_1^{(1)}(t) = \underline{m}_1^{(1)} = m_1^{(1)}(s)$$

and it is obvious that there exists $\gamma \in \Gamma$ with

$$t = \gamma(t_o) \in \Gamma(t_o) \quad .$$

Next , suppose that

$$B).\ m_1^{(1)}(t_o) \leqslant m_1^{(1)}(s) \quad .$$

Since $s \in \Theta$ we have

$$m_1^{(1)}(s) \leqslant L_1 = \sup \left\{ m_n^{(1)}(t_o)\ ;\ n \geqslant 1 \right\} ,$$

so there exists a unique $h \geqslant 1$ such that

$$m_1^{(h)}(t_o) \leqslant m_1^{(1)}(s) \quad \text{and} \quad m_1^{(h+1)}(t_o) \geqslant m_1^{(1)}(s) \ .$$

We define

$$\bar{m}_i^{(k)} = \begin{cases} m_1^{(1)}(s) & \text{if } i = 1 \text{ and } k \leq h \quad , \\ m_i^{(k)}(t_o) & \text{if } i \neq 1 \text{ or } k > h \quad . \end{cases}$$

Then $Af(s;1,1)$ holds with

$$t = \left\{(\bar{m}_i^{(k)})_{1 \leq i \leq k}\right\}_{1 \leq k < +\infty} \in \Omega \quad .$$

Now we prove (ii) . Choose $t \in \Gamma(t_o)$ satisfying the assertion $Af(s;n,j)$. Again , there are two possibilities :

A). $m_{j+1}^{(n)}(t) \geqslant m_{j+1}^{(n)}(s)$ or B). $m_{j+1}^{(n)}(t) \leqslant m_{j+1}^{(n)}(s)$

and we begin with the first one , so we suppose that

$$A).\ m_{j+1}^{(n)}(t) \geqslant m_{j+1}^{(n)}(s) \quad .$$

We continue in two steps .

(I_A) <u>We show that there exists</u> $h \geqslant 0$ <u>such that the following statement is true</u>

$$St_A(s;h) : m_{j+h}^{(n+h-1)}(t) \geqslant m_{j+1}^{(n)}(s) + 1 \ \underline{\text{and}}\ m_{j+h+1}^{(n+h)}(t) \leqslant m_{j+1}^{(n)}(s) \ .$$

Suppose the contrary holds . Then

$$m_{j+h}^{(n+h-1)}(t) \geqslant m_{j+1}^{(n)}(s) + 1 \quad \text{for every} \quad h \geqslant 0 \ .$$

Since $t \in \Gamma(t_o)$ there exists $h_o \geqslant 0$ such that

$$m_{j+h}^{(n+h-1)}(t) = m_{j+h}^{(n+h-1)}(t_o) \text{ for every } h \geqslant h_o.$$

By the definition of M_{n-j} we have

$$M_{n-j} = \inf_{h \in \mathbb{N}} m_{j+h}^{(n+h-1)}(t_o) \geqslant m_{j+1}^{(n)}(s) + 1 \quad .$$

On the other hand , $s \in \Theta$, hence

$$M_{n-j} \leqslant m_{j+1}^{(n)}(s) \quad .$$

This is a contradiction .

(II_A) <u>We show by induction on</u> $h \geqslant 0$ <u>that</u>

$$St_A(s;h) \Longrightarrow Af(s;n,j+1) \quad .$$

It is obvious that $St_A(s;0) \Longrightarrow Af(s;n,j+1)$. Suppose we have proved that $St_A(s;h-1) \Longrightarrow Af(s;n,j+1)$. From $St_A(s;h)$ we infer

$$(3) \qquad \begin{cases} m^{(n+h)}_{j+h}(t) \geqslant m^{(n+h-1)}_{j+h}(t) \geqslant m^{(n)}_{j+1}(s) + 1 \quad , \\ m^{(n+h)}_{j+h+1}(t) \leqslant m^{(n)}_{j+1}(s) \quad . \end{cases}$$

Since t satisfies $Af(s;n,j)$ we have

$$(4) \qquad m^{(n+h-2)}_{j+h}(t) \leqslant m^{(n-1)}_{j+1}(t) = m^{(n-1)}_{j+1}(s) \leqslant m^{(n)}_{j+1}(s).$$

We define

$$\underline{m}^{(k)}_{i.} = \begin{cases} m^{(k)}_{i}(t) \quad \text{if} \quad k \neq n+h-1 \quad \text{or if} \quad k = n+h-1 \quad \text{and} \quad i < j+h \ , \\ m^{(n)}_{j+1}(s) \quad \text{if} \quad k = n+h-1 \quad \text{and} \quad i = j+h \ , \\ \inf\left\{ m^{(k-1)}_{i-1}(t) \ , \ m^{(k+1)}_{i}(t) \right\} \quad \text{if} \quad k = n+h-1 \text{ and } i > j+h \ . \end{cases}$$

Using (3) and (4) we see that

$$t' = \left\{ (\underline{m}^{(k)}_{i})_{1 \leqslant i \leqslant k} \right\}_{1 \leqslant k < +\infty} \in \Omega$$

and it is obvious that there exists $\gamma \in \Gamma$ such that

$$t' = \gamma(t) \in \Gamma(t_o) \quad .$$

Then both $Af(s;n,j)$ and $St_A(s,h-1)$ are satisfied replacing t by t' . The induction hypothesis in (II_A) insures that $Af(s;n,j+1)$ holds .

Next , suppose that

$$B). \ m^{(n)}_{j+1}(t) \leqslant m^{(n)}_{j+1}(s) \quad .$$

We proceed again in two steps .

(I_B) <u>We show that there exists</u> $h \geqslant 0$ <u>such that the</u>

<u>following statement is true</u>

$$St_B(s;h) : m^{(n+h-1)}_{j+1}(t) \leqslant m^{(n)}_{j+1}(s) - 1 \quad \underline{\text{and}} \quad m^{(n+h)}_{j+1}(t) \geqslant m^{(n)}_{j+1}(s).$$

Indeed, since $s \in \Theta$, in the contrary case we would obtain the following contradiction :

$$m^{(n)}_{j+1}(s) \leqslant L_{j+1} = \sup_{h\in\mathbb{N}} m^{(n+h-1)}_{j+1}(t_o) = \sup_{h\in\mathbb{N}} m^{(n+h-1)}_{j+1}(t) \leqslant m^{(n)}_{j+1}(s) - 1 .$$

(II$_B$) <u>We show by induction on</u> $h \geqslant 0$ <u>that</u>

$$St_B(s;h) \Longrightarrow Af(s;n,j+1) .$$

Indeed, from $St_B(s;h)$ we infer

$$(5) \qquad \begin{cases} m^{(n+h)}_{j+2}(t) \leqslant m^{(n+h-1)}_{j+1}(t) \leqslant m^{(n)}_{j+1}(s) - 1 , \\ m^{(n+h)}_{j+1}(t) \geqslant m^{(n)}_{j+1}(s) , \end{cases}$$

and since t satisfies $Af(s;n,j)$ we have

$$(6) \qquad m^{(n+h-2)}_{j}(t) \geqslant m^{(n-1)}_{j}(t) = m^{(n-1)}_{j}(s) \geqslant m^{(n)}_{j+1}(s).$$

Putting

$$\bar{m}^{(k)}_{i} = \begin{cases} m^{(k)}_{i}(t) & \text{if } k \neq n+h-1 \text{ or if } k = n+h-1 \text{ and } i < j+1 , \\ m^{(n)}_{j+1}(s) & \text{if } k = n+h-1 \text{ and } i = j+1 , \\ \inf\left\{m^{(k-1)}_{i-1}(t) , m^{(k+1)}_{i}(t)\right\} & \text{if } k = n+h-1 \text{ and } i > j+1 , \end{cases}$$

and using (5) and (6) we obtain

$$t' = \left\{(\bar{m}^{(k)}_{i})_{1 \leqslant i \leqslant k}\right\}_{1 \leqslant k < +\infty} \in \Omega$$

which satisfies both $Af(s;n,j)$ and $St_B(s;h-1)$. Hence the induction hypothesis in (II$_B$) implies that $Af(s;n,j+1)$ holds.

The proof of (iii) is similar and we omit it .

We continue the proof of the Lemma . By (1) and (2) we have

$$\Gamma(t_0) \subset \Theta \subset \overline{\Gamma(t_0)}$$

and so

$$\omega = \overline{\Theta} \qquad (7)$$

But Θ is obviously closed in $\{t \in \Omega \,;\, n_0(t) = +\infty\}$ with respect to the relative topology , thus

$$\omega \cap \{t \in \Omega \,;\, n_0(t) = +\infty\} = \Theta \,,$$

which proves the first part of the second statement of the Lemma.

Consider now $s \in \Omega$ with $n_0(s) < +\infty$. Then $s \notin \Theta$. Therefore , $s \in \overline{\Theta}$ if and only if the set of all symbols

$$(m_j^{(n_0(s))}(t))_{1 \leq j \leq n_0(s)} ,$$

with

$$t \in \Theta \text{ and } m_j^{(k)}(t) = m_j^{(k)}(s) \text{ for } 1 \leq j \leq k < n_0(s) ,$$

is an infinite set . Hence $s \in \omega = \overline{\Theta}$ implies

$$\sup\left\{m_1^{(n_0(s))}(t) - m_{n_0(s)}^{(n_0(s))}(t) \,;\, t \in \Theta\right\} = +\infty$$

and so

$$L_1 - M_1 = +\infty .$$

Conversely , if $L_1 - M_1 = +\infty$ then it is clear that any point $s \in \Omega$ with $n_0(s) < +\infty$ and

$$L_j \geq m_j^{(n)}(s) \geq M_{n-j+1} \quad \text{for all } 1 \leq j \leq n < n_0(s)$$

belongs to $\overline{\Theta} = \omega$.

This completes the proof .

Q.E.D.

III.1.4. The next Lemma answers a natural converse question.

LEMMA . <u>For any given</u> $L_j \in \mathbb{Z} \cup \{+\infty\}$, $M_j \in \mathbb{Z} \cup \{-\infty\}$ $(j \in \mathbb{N})$ <u>such that</u>

$$+\infty \geqslant L_1 \geqslant L_2 \geqslant \dots \geqslant L_j \geqslant \dots \geqslant M_j \geqslant \dots \geqslant M_2 \geqslant M_1 \geqslant -\infty$$

<u>there exists a point</u> $t_o \in \Omega$ <u>with</u> $n_o(t_o) = +\infty$ <u>such that</u>

$$L_j = L_j(t_o) \quad , \quad M_j = M_j(t_o) \quad , \quad j \in \mathbb{N} \quad .$$

<u>Proof</u> . We distinguish three different situations :

1). <u>Suppose there is</u> $a \in \mathbb{Z}$ <u>with</u> $\inf_{j \in \mathbb{N}} L_j \geqslant a \geqslant \sup_{j \in \mathbb{N}} M_j$.

Then the point $t_o \in \Omega$ we are looking for can be defined as follows

$$m_j^{(2n)}(t_o) = \begin{cases} \inf \{L_j \;,\; a+n-j\} & \text{if } 1 \leqslant j \leqslant n \\ \sup \{M_{2n-j+1} \;,\; a+n-j+1\} & \text{if } n+1 \leqslant j \leqslant 2n \end{cases}$$

$$m_j^{(2n+1)}(t_o) = \begin{cases} \inf \{L_j \;,\; a+n-j+1\} & \text{if } 1 \leqslant j \leqslant n+1 \\ \sup \{M_{2n-j+2} \;,\; a+n-j+2\} & \text{if } n+2 \leqslant j \leqslant 2n+1 \end{cases}$$

2). <u>Suppose that</u> $\inf_{j \in \mathbb{N}} L_j = -\infty$, <u>so</u> $M_j = -\infty$ <u>for all</u> $j \in \mathbb{N}$.

Then the point $t_o \in \Omega$ we are looking for can be defined by

$$m_j^{(n)}(t_o) = \inf \{L_j \;,\; n-j\} \quad , \quad 1 \leqslant j \leqslant n < +\infty .$$

3). <u>Suppose that</u> $\sup_{j \in \mathbb{N}} M_j = +\infty$, <u>so</u> $L_j = +\infty$ <u>for all</u> $j \in \mathbb{N}$.

Then the point $t_o \in \Omega$ we are looking for can be defined by

$$m_j^{(n)}(t_o) = \sup \{M_{n-j+1} \;,\; -j+1\}, \quad 1 \leqslant j \leqslant n < +\infty .$$

Since $\inf_{j \in \mathbb{N}} L_j \geqslant \sup_{j \in \mathbb{N}} M_j$, these three cases cover all the possible situations and the Lemma is proved .

Q.E.D.

It is obvious that for any given integers

$$L_1 \geqslant L_2 \geqslant \dots \geqslant L_{n_0-1} \quad , \quad n_0 < +\infty \quad ,$$

there is a point $t_0 \in \Omega$, with $n_0(t_0) = n_0$, such that

$$m_j^{(n_0-1)}(t_0) = L_j \quad \text{for all} \quad 1 \leqslant j < n_0 \quad .$$

III.1.5. Thus , taking into account Theorem I.2.9. , Theorem II.1.5. , the remarks in Section III.1.1. and the preceding Lemmas , we obtain the following

THEOREM . <u>The primitive spectrum of the C^*- algebra</u> $A(U(\infty))$ <u>can be identified with the set of all the symbols</u>

$$\xi = \left\{L_j(\xi) \ , \ M_j(\xi)\right\}_{1 \leqslant j < n_0(\xi)}$$

<u>where either</u> $n_0(\xi) = +\infty$ <u>and</u> , <u>for all</u> $1 \leqslant j < +\infty$, <u>we have</u>

$$\mathbb{Z} \cup \{+\infty\} \ni L_j(\xi) \geqslant L_{j+1}(\xi) \geqslant M_{j+1}(\xi) \geqslant M_j(\xi) \in \mathbb{Z} \cup \{-\infty\} \quad ,$$

<u>or</u> $n_0(\xi) \in \mathbb{N}$ <u>and, for all</u> $1 \leqslant j < n_0(\xi)$, <u>we have</u>

$$\mathbb{Z} \ni M_{n_0-j}(\xi) = L_j(\xi) \geqslant L_{j+1}(\xi) = M_{n_0-j-1}(\xi) \quad .$$

<u>Namely , if</u> ρ <u>is a factor representation of</u> $U(\infty)$ (<u>or of some</u> $U(k)$) , <u>then the kernel of</u> π_ρ <u>corresponds to the symbol</u>

$$L_j = \sup\left\{\sup\left\{m_j^{(n)} ; \ n \geqslant j\right\}\right\} \quad , \quad M_j = \inf\left\{\inf\left\{m_{n-j+1}^{(n)} ; \ n \geqslant j\right\}\right\}$$

<u>where the first</u> sup <u>and the first</u> inf <u>are taken over all signatures</u> $(m_1^{(n)}, \dots , m_n^{(n)}) \in \widehat{U(n)}$ <u>which appear in</u> $\rho \,|\, U(n)$.

The points $\xi \in \mathrm{Prim}(A(U(\infty)))$ with $n_0(\xi) = +\infty$ correspond to factor representations of $U(\infty)$, while the points ξ with $n_0(\xi) = n_0 \in \mathbb{N}$ correspond to factor representations of $U(n_0-1)$.

It is easy to describe <u>the topology on the space</u>

$$\mathrm{Prim}(A(U(\infty))) \ .$$

Thus, consider $E \subset \mathrm{Prim}(A(U(\infty)))$ and $\xi_0 \in \mathrm{Prim}(A(U(\infty)))$.
Then <u>we have</u>

$$\xi_0 \in \overline{E}$$

<u>if and only if any of the following conditions holds</u> :

(i) $\xi_0 \in E$;

(ii) $n_0(\xi_0) = +\infty$ <u>and</u>, <u>for all</u> $1 \leqslant j < +\infty$,

$$L_j(\xi_0) \leqslant \sup\{L_j(\xi) \ ; \xi \in E\} \ , \quad M_j(\xi_0) \geqslant \inf\{M_j(\xi) \ ; \xi \in E\} \ ;$$

(iii) $n_0(\xi_0) = n_0 < +\infty$, $\sup\left\{L_1(\xi) - M_1(\xi) \ ; \ \xi \in E\right\} = +\infty$

<u>and, for all</u> $1 \leqslant j < n_0$,

$$\sup\left\{L_j(\xi) \ ; \ \xi \in E\right\} \geqslant L_j(\xi_0) = M_{n_0 - j}(\xi_0) \geqslant \inf\left\{M_{n_0 - j}(\xi) \ ; \ \xi \in E\right\} \ .$$

For example, the one point set $\{\xi_\infty\} \subset \mathrm{Prim}(A(U(\infty)))$, where

$$L_j(\xi_\infty) = +\infty \ , \quad M_j(\xi_\infty) = -\infty \ \text{for all} \ 1 \leqslant j < +\infty,$$

is everywhere dense .

§ 2 <u>Direct limits of irreducible representations</u>

The direct limits of irreducible representations of the U(n)'s provide us with a wide class of irreducible representations of $U(\infty)$. In particular , every primitive ideal of $A(U(\infty))$ which corresponds to $U(\infty)$ is the kernel of at least one such representation .

Among these representations there are also the representations considered by I.E. Segal ([30]) and A.A.Kirillov ([21]) .

Consider a point $t \in \Omega = \Omega(U(\infty))$ with $n_0(t) = +\infty$. Thus ,

$$t = (\rho_1 \prec \rho_2 \prec \dots \prec \rho_n \prec \dots)$$

where $\rho_n \in \widehat{U(n)}$, $n \in \mathbb{N}$.

III.2.1. Let μ be a completely atomic Γ- quasi-invariant probability measure concentrated on the Γ- orbit $\Gamma(t)$. Then , for all $\gamma \in \Gamma$ we have

$$\mu(\{\gamma(t)\}) > 0$$

and , for all Borel sets $B \subset \Omega$ we have

$$\mu(B) = \sum_{s \in \Gamma(t) \cap B} \mu(\{s\}) \quad .$$

Clearly , μ is ergodic and therefore the representation ρ_μ is irreducible (see I.3.16.) .

Moreover , the kernel of ρ_μ corresponds to $\overline{\Gamma(t)}$ (see I.2.9.) .

III.2.2. On the other hand , since $\rho_n \prec \rho_{n+1}$, there are isometric embeddings

$$i_n : H_{\rho_n} \longrightarrow H_{\rho_{n+1}} \quad ,$$

such that

$$(\rho_{n+1}|U(n)) \circ i_n = i_n \circ \rho_n \quad .$$

Moreover , since $[\rho_{n+1} : \rho_n] = 1$, the i_n's are unique up to a scalar factor of module 1 .

On the completion H_t of the direct limit of the H_{ρ_n} 's following the i_n's there is a natural representation of $U(\infty)$. It is easy to see that the representations corresponding to two different choices of the i_n's are unitarily equivalent . Therefore , we may denote this direct limit representation by ρ_t .

III.2.3. Any two representations $\rho^{(j)}$ of $U(\infty)$ on Hilbert spaces $H^{(j)}$, $j = 1,2$, such that the subspaces

$$\bigcap_{n=0}^{\infty} \rho^{(j)}(p_{\rho_n})\, H^{(j)}$$

are one dimensional and cyclic , are unitarily equivalent .

Indeed , if

$$\xi^{(j)} \in \bigcap_{n=0}^{\infty} \rho^{(j)}(p_{\rho_n})\, H^{(j)} \quad , \quad \|\xi^{(j)}\| = 1 \quad ,$$

then the functions of positive type determined by $\xi^{(1)}$ and $\xi^{(2)}$ on $U(\infty)$ are equal , as can be easily seen considering the restrictions to the various $U(n)$.

III.2.4. From the above remark we infer :

<u>The representations</u> ρ_μ <u>and</u> ρ_t <u>are unitarily equivalent</u>.

In particular ,

<u>The representation</u> ρ_t <u>is irreducible and the associated primitive ideal of</u> $A(U(\infty))$ <u>corresponds to the closure of</u> Γ- <u>orbit</u> $\overline{\Gamma(t)}$.

Moreover ,

Two representations ρ_t *and* $\rho_{t'}$ *are unitarily equivalent if and only if* $t' = \gamma(t)$ *for some* $\gamma \in \Gamma$.

III.2.5. In the general case of direct limits of compact groups , the direct limits of irreducible representations of the G_n ' s are still irreducible representations of G_∞ . Also , for such a representation one can choose a system of matrix units such that the representation be unitarily equivalent to ρ_μ with μ concentrated on the corresponding Γ - orbit .

CHAPTER IV TYPE III FACTOR REPRESENTATIONS OF $U(\infty)$ IN ANTISYMMETRIC TENSORS

We shall study some representations of $U(\infty)$ whose restrictions to the $U(n)$'s contain only irreducible representations in antisymmetric tensors , i.e. representations with signatures of the form

$$(1,\ldots,1,0,\ldots,0) \quad .$$

IV.1. The notations and the results contained in Section III.1.1. will be used without any further reference.

Consider the set $\omega \subset \Omega$ consisting of all symbols

$$(\rho_1 \prec \rho_2 \prec \ \ldots \ \prec \rho_n \prec \ldots)$$

with $\rho_n \in \widehat{U(n)}$ of the form

$$\rho_n = (\underbrace{1,\ldots,1}_{k_n\text{-times}},\underbrace{0,\ldots\ldots,0}_{(n-k_n)\text{-times}})$$

Clearly , ω can be identified with the set of all sequences $\{k_n\}_{n\in\mathbb{N}}$ of positive integers enjoing the properties

$$k_1 \in \{0,1\} \quad , \quad k_{n+1} - k_n \in \{0,1\} \quad .$$

The map

$$\{k_n\}_{n\in\mathbb{N}} \longmapsto \{k_1 , k_2 - k_1 , \ldots , k_{n+1} - k_n , \ldots\}$$

allow us to identify ω with the product set

$$\{0,1\}^{\aleph_0} \quad .$$

It is easy to see that by this identification the topology of ω corresponds to the product topology of discrete topologies on $\{0,1\}^{\aleph_0}$.

The set ω is a Γ- orbit whose closure corresponds (see III.1.5.) to the upper signature $\left\{L_j = 1 \; ; \; j \in \mathbb{N}\right\}$ and to the lower signature $\left\{M_j = 0 \; ; \; j \in \mathbb{N}\right\}$. Restricting the transformations in Γ to ω we get a transformation group Γ_ω on ω. In order to describe this group in the identification

$$\omega = \{0,1\}^{\aleph_0}$$

we fix $n \in \mathbb{N}$ and a permutation σ of the set $\{0,1\}^n$ such that

$$\sigma(\alpha_1,\ldots,\alpha_n) = (\beta_1,\ldots,\beta_n) \Longrightarrow \sum_{i=1}^{n} \alpha_i = \sum_{i=1}^{n} \beta_i \quad ,$$

and we define a transformation $\gamma_{n,\sigma}$ on ω by

$$\gamma_{n,\sigma}(\alpha_1,\ldots,\alpha_n,\alpha_{n+1},\ldots) = (\sigma(\alpha_1,\ldots,\alpha_n),\alpha_{n+1},\ldots) \quad .$$

Then it is easy to see that <u>the group Γ_ω consists of all transformations $\gamma_{n,\sigma}$ where $n \in \mathbb{N}$ and σ is a permutation of the set $\{0,1\}^n$ which preserves the sum of the components of the elements</u>.

We remark that the set $\Gamma_{n_0,\omega}$ of all transformations $\gamma_{n,\sigma}$ with $n = n_0$ is a subgroup of Γ_ω.

We consider on $\omega = \{0,1\}^{\aleph_0}$ the product measures

$$\mu = \bigotimes_{n=1}^{\infty} \mu_n$$

where each μ_n is a probability measure on $\{0,1\}$:

$$\mu_n(\{0\}) = p_n^{(0)} \quad , \quad \mu_n(\{1\}) = p_n^{(1)} \quad ,$$

$$0 < p_n^{(0)} < 1 \quad , \quad 0 < p_n^{(1)} < 1 \quad ,$$

$$p_n^{(0)} + p_n^{(1)} = 1 \quad .$$

It is obvious that any such measure μ on ω is Γ_ω -

quasi-invariant .

The representations we shall study are the representations π_μ, ρ_μ of $A(U(\infty))$ which correspond to μ by the general considerations of Chapter I § 3 . Therefore we are interested in the Γ_ω - ergodicity of μ , in the μ - measurability of Γ_ω and in the equivalence of two measures of the above type .

IV.2. Concerning the ergodicity , the main instrument is Proposition I.3.14, which was obtained as a consequence of the Powers-Bratteli theorem . Using this Proposition , it follows that the measure μ is Γ_ω - ergodic if and only if

for every $f \in C(\omega)$ and every $\varepsilon > 0$ there is $r \in \mathbb{N}$ such that for every $\Gamma_{r,\omega}$- invariant $g \in C(\omega)$ we have

$$\left| \int fg \, d\mu - \left(\int f \, d\mu\right)\left(\int g \, d\mu\right)\right| < \varepsilon \, \|g\|_{C(\omega)} \quad . \tag{1}$$

It easy to see that it suffices to verify this condition only for functions $f \in C(\omega)$ of the form

$$f((\beta_j)_{j=1}) = \begin{cases} 0 & \text{if} \quad (\beta_j)_{j=1}^n \neq (\alpha_j)_{j=1}^n \\ 1 & \text{if} \quad (\beta_j)_{j=1}^n = (\alpha_j)_{j=1}^n \end{cases} \quad ; \quad (\beta_j)_{j=1}^\infty \in \omega \quad ,$$

where $n \in \mathbb{N}$, $(\alpha_j)_{j=1}^n \in \{0,1\}^n$ are arbitrary but fixed .

Moreover , we may suppose that the function $g \in C(\omega)$ depends only on a finite (but non-fixed) number $N \in \mathbb{N}$ of components of its argument .

Thus , consider $n < r < N$. Since the function g is $\Gamma_{r,\omega}$- invariant , there is a function φ such that

$$g((\beta_j)_{j=1}^{\infty}) = \varphi(\beta_1+\dots+\beta_r \,;\, \beta_{r+1}, \dots, \beta_N), \quad (\beta_j)_{j=1}^{\infty} \in \omega .$$

Let us denote

$$D_k = \left\{ (\beta_j)_{j=1}^{\infty} \in \omega \;;\; \sum_{j=1}^{r} \beta_j = k \right\}, \quad 0 \leq k \leq r ,$$

$$D_k' = D_k \cap \left\{ (\beta_j)_{j=1}^{\infty} \in \omega \;;\; (\beta_j)_{j=1}^{n} = (\alpha_j)_{j=1}^{n} \right\}, \quad 0 \leq k \leq r .$$

Then

$$\int f \, d\mu = \prod_{j=1}^{n} p_j^{(\alpha_j)}$$

$$\int g \, d\mu = \sum_{k=0}^{n} \sum_{(\gamma_j)_{j=1}^{N-r} \in \{0,1\}^{N-r}} \varphi(k, \gamma_1, \dots, \gamma_{N-r}) \, \mu(D_k) \prod_{j=1}^{N-r} p_{r+j}^{(\gamma_j)}$$

$$\int fg \, d\mu = \sum_{k=0}^{n} \sum_{(\gamma_j)_{j=1}^{N-r} \in \{0,1\}^{N-r}} \varphi(k, \gamma_1, \dots, \gamma_{N-r}) \, \mu(D_k') \prod_{j=1}^{N-r} p_{r+j}^{(\gamma_j)}$$

and so , the relation (1) becomes

$$\left| \sum_{(\gamma_j)_{j=1}^{N-r} \in \{0,1\}^{N-r}} \left(\prod_{j=1}^{N-r} p_{r+j}^{(\gamma_j)} \right) \left(\sum_{k=0}^{r} \varphi(k, \gamma_1, \dots, \gamma_{N-r}) \left(\mu(D_k') - \mu(D_k) \prod_{i=1}^{n} p_i^{(\alpha_i)} \right) \right) \right| \leq \varepsilon \|\varphi\|$$

This last relation is implied by the following one :

$$(2) \qquad \sum_{k=0}^{r} \left| \mu(D_k') - \mu(D_k) \prod_{i=1}^{n} p_i^{(\alpha_i)} \right| \leq \varepsilon$$

Putting

$$k_o = \sum_{i=1}^{n} \alpha_i ,$$

$$D_k(j,h) = \left\{ (\beta_i)_{i=1}^{\infty} \in \omega \;;\; \sum_{i=j}^{h} \beta_i = k \right\}, \quad j \leq h ,$$

we have

$$\mu(D_k) = \sum_{j=0}^{n} \mu(D_j(1,n))\,\mu(D_{k-j}(n+1,r)) \quad ,$$

$$\mu(D_k') = \mu(D_{k-k_0}(n+1,r)) \prod_{i=1}^{n} p_i^{(\alpha_i)} \quad .$$

Thus , in order to prove the Γ_ω - ergodicity of μ , it is sufficient to show that

$$\lim_{r\to\infty} \left(\prod_{i=1}^{n} p_i^{(\alpha_i)}\right) \sum_{k=0}^{r} \left| \mu(D_{k-k_0}(n+1,r)) - \sum_{j=0}^{n} \mu(D_j(1,n))\,\mu(D_{k-j}(n+1,r)) \right| = 0.$$

Since

$$\sum_{j=0}^{n} \mu(D_j(1,n)) = 1 \quad ,$$

it is sufficient to show that

(3) $$\lim_{r\to\infty} \sum_{k=0}^{\infty} \left| \mu(D_k(n+1,r)) - \mu(D_{k+1}(n+1,r)) \right| = 0 \quad .$$

IV.3. For technical reasons we consider the power series

$$P(z) = \sum_{k\in\mathbb{Z}} c_k\, z^k \quad \text{with} \quad \sum_{k\in\mathbb{Z}} |c_k| < +\infty$$

and we define

$$\|P(z)\| = \sum_{k\in\mathbb{Z}} |c_k| \quad .$$

It is easy to see that $\|\cdot\|$ is a norm and

(4) $$\|P_1(z)\,P_2(z)\| \leqslant \|P_1(z)\|\,\|P_2(z)\| \quad ,$$

(5) $$\|z^n\,P(z)\| = \|P(z)\| \quad .$$

We remark that $\mu(D_k(n+1,r))$ is exactly the coefficient c_k in the development

$$\prod_{i=n+1}^{r} (p_i^{(0)} + z\,p_i^{(1)}) = \sum_{k\in\mathbb{Z}} c_k\, z^k$$

and that for any $I \subset \mathbb{N}$ we have

(6) $$\left\| \prod_{i \in I} (p_i^{(0)} + z\, p_i^{(1)}) \right\| = 1 \quad .$$

Finally, the sufficient condition (3) for the Γ_ω - ergodicity of μ is rewritten as follows :

(7) $$\lim_{r \to \infty} \left\| (1 - z) \prod_{i=n+1}^{r} (p_i^{(0)} + z\, p_i^{(1)}) \right\| = 0 \quad .$$

IV.4. PROPOSITION . _If_ $\sum_{i=1}^{\infty} \min \{ p_i^{(0)} , p_i^{(1)} \} = +\infty$ _then the measure_ μ _is_ Γ_ω - _ergodic_ .

Proof . By the assumption in the statement we distinguish the following three cases :

A). there is a sequence $n < i_1 < i_2 < \dots$ such that

$$\lim_{s \to \infty} p_{i_s}^{(o)} = p^{(o)} \quad , \quad 0 < p^{(o)} < 1 \quad ;$$

B). there is a sequence $n < i_1 < i_2 < \dots$ such that

$$\lim_{s \to \infty} p_{i_s}^{(0)} = 0 \quad , \quad \sum_{s=1}^{\infty} p_{i_s}^{(0)} = +\infty \quad ;$$

C). there is a sequence $n < i_1 < i_2 < \dots$ such that

$$\lim_{s \to \infty} p_{i_s}^{(1)} = 0 \quad , \quad \sum_{s=1}^{\infty} p_{i_s}^{(1)} = +\infty \quad .$$

IV.5. _Case A)._ For graphical convenience we put

$$p^{(0)} = \lambda \quad , \quad 1 - p^{(0)} = \nu \quad .$$

Then we have

$$\overline{\lim_{r \to \infty}} \left\| (1 - z) \prod_{h=n+1}^{r} (p_h^{(0)} + z\, p_h^{(1)}) \right\|$$

$$\leqslant \overline{\lim_{m \to \infty}} \left\| (1 - z) \prod_{s=1}^{m} (p_{i_s}^{(0)} + z\, p_{i_s}^{(1)}) \right\|$$

$$\leqslant \overline{\lim_{m_2 \to \infty}} \left(\overline{\lim_{m_1 \to \infty}} \left\| (1 - z) \prod_{s=m_1}^{m_1+m_2} (p_{i_s}^{(0)} + z\, p_{i_s}^{(1)}) \right\| \right)$$

$$= \overline{\lim_{m \to \infty}} \left\| (1 - z)(\lambda + \nu z)^m \right\|$$

Since $\lim_{m \to \infty} \nu^m = 0$, in order to prove (7) it is sufficient to show that

$$(8) \qquad \lim_{m \to \infty} \sum_{k=0}^{m} C_m^k \lambda^k \nu^{m-k} \left| 1 - \frac{\lambda}{\nu} \frac{m - k}{k + 1} \right| = 0 \ ,$$

where C_m^k stands for the binomial coefficient .

Now fix $\varepsilon > 0$ and consider the sets

$A(m)$, $B(m)$, $C(m)$, $D(m)$, $E(m)$, $F(m)$,

whose elements are the integers k , $0 \leqslant k \leqslant m$, such that the quotient

$$\frac{\lambda}{\nu} \frac{m - k}{k + 1}$$

belongs respectively to the intervals

$(1+2\varepsilon, +\infty)$, $(1+\varepsilon, 1+2\varepsilon]$, $(1, 1+\varepsilon]$, $(\frac{1}{1+\varepsilon}, 1]$, $(\frac{1}{1+2\varepsilon}, \frac{1}{1+\varepsilon}]$, $[0, \frac{1}{1+2\varepsilon}]$.

There exist a , b , $c > 0$ and $N_o \in \mathbb{N}$ such that for $m > N_o$ we have

$$\text{card } A(m) < a\, m \ , \quad \text{card } F(m) < a\, m \ ,$$

$$\text{card } B(m) > b\, m \ , \quad \text{card } E(m) > b\, m \ ,$$

$$\text{card } C(m) > c\, m \ , \quad \text{card } D(m) > c\, m \ .$$

Moreover , for any $k_1 \in A(m)$ and any $k_2 \in C(m)$ we have

$$C_m^{k_1} \lambda^{k_1} \nu^{m-k_1} \leqslant (1 + \varepsilon)^{-bm} C_m^{k_2} \lambda^{k_2} \nu^{m-k_2} \ ;$$

also , for any $k_1 \in F(m)$ and any $k_2 \in D(m)$ we have

$$C_m^{k_1} \lambda^{k_1} \nu^{m-k_1} \leqslant (1 + \varepsilon)^{-bm} C_m^{k_2} \lambda^{k_2} \nu^{m-k_2} \ .$$

Thus

$$\sum_{k \in A(m) \cup F(m)} C_m^k \lambda^k \nu^{m-k} \leqslant$$

$$\leqslant \frac{a}{c}(1+\varepsilon)^{-bm} \sum_{k \in C(m)} C_m^k \lambda^k \nu^{m-k} + \frac{a}{c}(1+\varepsilon)^{-bm} \sum_{k \in D(m)} C_m^k \lambda^k \nu^{m-k}$$

$$\leqslant 2\,\frac{a}{c}\,(1+\varepsilon)^{-bm} \quad .$$

On the other hand , for $k \in B(m) \cup C(m) \cup D(m) \cup E(m)$ we have

$$\left| 1 - \frac{\lambda}{\nu} \frac{m-k}{k+1} \right| \leqslant 2\varepsilon \quad .$$

Since $\sum_{k=0}^{m} C_m^k \lambda^k \nu^{m-k} = 1$, it follows that

$$\left\| (1-z)(\lambda+\nu z)^m \right\| \leqslant \left\| (1-z) \sum_{k \in A(m) \cup F(m)} C_m^k \lambda^k \nu^{m-k} z^{m-k} \right\| +$$

$$+ \left\| (1-z) \sum_{k \in B(m) \cup C(m) \cup D(m) \cup E(m)} C_m^k \lambda^k \nu^{m-k} z^{m-k} \right\|$$

$$\leqslant 2 \sum_{k \in A(m) \cup F(m)} C_m^k \lambda^k \nu^{m-k} +$$

$$+ \sum_{k \in B(m) \cup C(m) \cup D(m) \cup E(m)} C_m^k \lambda^k \nu^{m-k} \left| 1 - \frac{\lambda}{\nu} \frac{m-k}{k+1} \right|$$

$$\leqslant 4\,\frac{a}{c}\,(1+\varepsilon)^{-bm} + 2\varepsilon \quad .$$

Hence

$$\lim_{m \to \infty} \left\| (1-z)(\lambda+\nu z)^m \right\| = 0$$

and the measure μ on ω is Γ_ω - ergodic .

IV.6. <u>Case B).</u> We remark that

$$\left\| (1-z) \prod_{i=n+1}^{r} (p_i^{(0)} + z\,p_i^{(1)}) \right\| = \left\| (1-z^{-1}) \prod_{i=n+1}^{r} (p_i^{(0)} + z^{-1} p_i^{(1)}) \right\|$$

$$= \left\| (1 - z) \prod_{i=n+1}^{r} (z\, p_i^{(o)} + p_i^{(1)}) \right\| .$$

Therefore Case B). reduces to Case C).

IV.7. Case C). We shall need the following two lemmas .

LEMMA 1. Let $\lambda_1^{(n)}, \ldots, \lambda_{N(n)}^{(n)}$ be positive such that

$$\lim_{n\to\infty} \sum_{j=1}^{N(n)} \lambda_j^{(n)} = \lambda > 0 \quad \text{and} \quad \lim_{n\to\infty} \max_{1 \leq j \leq N(n)} \lambda_j^{(n)} = 0 .$$

Then

$$\lim_{n\to\infty} \left\| \prod_{j=1}^{N(n)} (1 + \lambda_j^{(n)} z) - e^{\lambda z} \right\| = 0$$

Proof of Lemma 1 . There is $n_0 \in \mathbb{N}$ such that for $n \geq n_0$ we have

$$\sum_{j=1}^{N(n)} \lambda_j^{(n)} < 2\lambda \quad \text{and} \quad \max_{1 \leq j \leq N(n)} \lambda_j^{(n)} < (1 + 2\lambda)^{-1}$$

Then the following inequalities are easily verified :

$$\left\| \prod_{j=1}^{N(n)} e^{\lambda_j^{(n)} z} - \prod_{j=1}^{N(n)} (1 + \lambda_j^{(n)} z) \right\| \leq$$

$$\leq \left\| \prod_{j=1}^{N(n)} (1 - \lambda_j^{(n)} z)^{-1} - \prod_{j=1}^{N(n)} (1 + \lambda_j^{(n)} z) \right\| \leq$$

$$\leq \left\| \prod_{j=1}^{N(n)} (1 - (\lambda_j^{(n)} z)^2)^{-1} - 1 \right\| \cdot \left\| \prod_{j=1}^{N(n)} (1 + \lambda_j^{(n)} z) \right\| \leq$$

$$\leq \left\| (1 - z^2 \sum_{j=1}^{N(n)} (\lambda_j^{(n)})^2)^{-1} - 1 \right\| \cdot e^{2\lambda} \leq$$

$$\leq \left[(1 - 2\lambda \max_{1 \leq j \leq N(n)} \lambda_j^{(n)})^{-1} - 1 \right] \cdot e^{2\lambda} ,$$

whence

$$\lim_{n\to\infty} \left\| \prod_{j=1}^{N(n)} e^{\lambda_j^{(n)} z} - \prod_{j=1}^{N(n)} (1 + \lambda_j^{(n)} z) \right\| = 0 .$$

Since

$$\lim_{n\to\infty} \left\| e^{z \sum_{j=1}^{N(n)} \lambda_j^{(n)}} - e^{z\lambda} \right\| = 0 ,$$

the Lemma follows .

LEMMA 2 . $\lim_{\lambda\to\infty} e^{-\lambda} \| (1 - z)\, e^{\lambda z} \| = 0$.

Proof of Lemma 2 . The proof is based on arguments similar to those used in Case A) to prove (8) , so we shall be brief in details .

Thus , fix $0 < \varepsilon < 1/2$ and consider the sets

$$A(\lambda) , \quad B(\lambda) , \quad C(\lambda) , \quad D(\lambda) , \quad E(\lambda) , \quad F(\lambda)$$

whose elements are the positive integers k such that the quotient

$$\frac{\lambda}{k+1}$$

belongs respectively to the intervals

$$[1+2\varepsilon,+\infty) , \ [1+\varepsilon,1+2\varepsilon) , \ [1,1+\varepsilon) , \ [\tfrac{1}{1+\varepsilon},1) , \ [\tfrac{1}{1+2\varepsilon},\tfrac{1}{1+\varepsilon}) , \ (0,\tfrac{1}{1+2\varepsilon}]$$

There exist $a > 0$ and $\lambda_0 > 0$ such that for $\lambda \geqslant \lambda_0$ we have

$$\text{card } B(\lambda) > a\lambda , \quad \text{card } E(\lambda) > a\lambda ,$$

$$\text{card } C(\lambda) > 0 , \quad \text{card } D(\lambda) > 0 .$$

Then

$$\sum_{k\in A(\lambda)\cup F(\lambda)} \frac{\lambda^k}{k!} \leqslant (1+\varepsilon)^{-a\lambda} \sum_{j=0}^{\infty} (1+\varepsilon)^{-j} \sum_{k\in C(\lambda)\cup D(\lambda)} \frac{\lambda^k}{k!}$$

and so

$$\lim_{\lambda\to\infty} e^{-\lambda}\left[\sum_{k\in A(\lambda)\cup F(\lambda)} \frac{\lambda^k}{k!}\right] = 0.$$

Finally ,

$$\overline{\lim_{\lambda\to\infty}}\; e^{-\lambda}\left\|(1 - z)\, e^{\lambda z}\right\| =$$

$$= \overline{\lim_{\lambda\to\infty}}\; e^{-\lambda}\left\|(1 - z)\sum_{k\in B(\lambda)\cup C(\lambda)\cup D(\lambda)\cup E(\lambda)} \frac{\lambda^k}{k!} z^k\right\| =$$

$$= \overline{\lim_{\lambda\to\infty}}\; e^{-\lambda}\sum_{k\in B(\lambda)\cup C(\lambda)\cup D(\lambda)\cup E(\lambda)} \frac{\lambda^k}{k!}\left|1 - \frac{\lambda}{k+1}\right| \leq$$

$$\leq 2\varepsilon$$

and this proves the Lemma .

Let us now return to <u>the proof of the Proposition</u> IV.4. <u>in Case</u> C). By Lemma 2 there is $\lambda > 0$ such that

$$e^{-\lambda}\left\|(1 - z)\, e^{\lambda z}\right\| < \varepsilon \quad .$$

Owing to the assumption in Case C) and to Lemma 1 , we find indices

$$n < j_1 < j_2 < \dots < j_p$$

such that

$$\left\| e^{-\lambda} e^{\lambda z} - \prod_{s=1}^{p} \left(p_{j_s}^{(o)} + z\, p_{j_s}^{(1)}\right)\right\| < \varepsilon \quad .$$

We infer

$$\overline{\lim_{r\to\infty}}\left\|(1 - z)\prod_{h=n+1}^{r} \left(p_h^{(o)} + z\, p_h^{(1)}\right)\right\| \leq$$

$$\leq \left\|(1 - z)\prod_{s=1}^{p} \left(p_{j_s}^{(o)} + z\, p_{j_s}^{(1)}\right)\right\| \leq$$

$$\leqslant \left\| (1 - z)\left(e^{-\lambda} e^{\lambda z} - \prod_{s=1}^{p} (p_{j_s}^{(0)} + z\, p_{j_s}^{(1)})\right)\right\| + \left\| (1 - z) e^{-\lambda} e^{\lambda z} \right\|$$

$$\leqslant 2\varepsilon + \varepsilon = 3\varepsilon .$$

Hence

$$\lim_{r \to \infty} \left\| (1 - z) \prod_{h=n+1}^{r} (p_h^{(0)} + z\, p_h^{(1)}) \right\| = 0$$

and the measure μ on ω is Γ_ω - ergodic .

This ends the proof of Proposition IV.4.

IV.8. Concerning the μ - measurability of Γ_ω we have

PROPOSITION . _If the sequence_ $\{p_i^{(0)}\}_{i=1}^{\infty}$ _has two different limit points_ $0 < p < p' < 1$, _then the group_ Γ_ω _is non-measurable with respect to_ μ .

Proof . By the assumption we can find a sequence

$$i_1 < i_2 < \dots < i_n < \dots$$

such that

$$\lim_{k \to \infty} p_{i_{2k}}^{(0)} = p \quad \text{and} \quad \lim_{k \to \infty} p_{i_{2k-1}}^{(0)} = p' .$$

We consider the transformations $\gamma_k \in \Gamma_\omega$ defined as follows :

$$\gamma_k((\alpha_j)_{j=1}^{\infty}) = (\beta_j)_{j=1}^{\infty}$$

where

a) if $\alpha_{2^k+2s-1} = \alpha_{2^k+2s}$ for every $1 \leqslant s \leqslant 2^{k-1}$ we put

$$(\beta_j)_{j=1}^{\infty} = (\alpha_j)_{j=1}^{\infty}$$

b) in the contrary case , let s_o be the smallest s with

$1 \leq s \leq 2^{k-1}$ and $\alpha_{2^k+2s-1} \neq \alpha_{2^k+2s}$; then we put

$$\beta_j = \alpha_j \text{ for all } j \in \mathbb{N} \setminus \{2^k+2s_0-1 \,,\, 2^k+2s_0\} \,,$$

$$\beta_{2^k+2s_0-1} = \alpha_{2^k+2s_0} \,,$$

$$\beta_{2^k+2s_0} = \alpha_{2^k+2s_0-1} \,.$$

Let us further denote

$$\theta = \sup_{k\in\mathbb{N}} \left\{ p^{(o)}_{i_{2k-1}} \, p^{(o)}_{i_{2k}} + p^{(1)}_{i_{2k-1}} \, p^{(1)}_{i_{2k}} \right\} \,,$$

$$M = \sup_{k\in\mathbb{N}} \max \left\{ \frac{p^{(o)}_{i_{2k-1}} \, p^{(1)}_{i_{2k}}}{p^{(1)}_{i_{2k-1}} \, p^{(o)}_{i_{2k}}} \,,\, \frac{p^{(1)}_{i_{2k-1}} \, p^{(o)}_{i_{2k}}}{p^{(o)}_{i_{2k-1}} \, p^{(1)}_{i_{2k}}} \right\} \,,$$

$$\varepsilon = \inf_{k\in\mathbb{N}} \min \left\{ \left| \frac{p^{(o)}_{i_{2k-1}} \, p^{(1)}_{i_{2k}}}{p^{(1)}_{i_{2k-1}} \, p^{(o)}_{i_{2k}}} - 1 \right| \,,\, \left| \frac{p^{(1)}_{i_{2k-1}} \, p^{(o)}_{i_{2k}}}{p^{(o)}_{i_{2k-1}} \, p^{(1)}_{i_{2k}}} - 1 \right| \right\} \,.$$

The assumptions insure that

$$0 < \theta < 1 \quad , \quad 0 < M < +\infty \quad , \quad \varepsilon > 0 \quad .$$

Denoting

$$\Delta_k = \{ \alpha \in \omega \,;\, \gamma_k(\alpha) = \alpha \} \,,$$

we see that

$$\mu(\Delta_k) \leq \theta^{2^{k-1}} \quad \text{so} \quad \sum_{k=1}^{\infty} \mu(\Delta_k) < +\infty .$$

Now , for $\alpha \in \Delta_k$ we have

$$\frac{d\,\mu^{\gamma_k}}{d\mu}(\alpha) = 1$$

and for $\alpha \in \omega \setminus \Delta_k$ we have

$$\frac{d\mu^{\gamma_k}}{d\mu}(\alpha) \in \left\{ \frac{p^{(o)}_{i_{2k-1}}\; p^{(1)}_{i_{2k}}}{p^{(1)}_{i_{2k-1}}\; p^{(o)}_{i_{2k}}} \; ; \; k \in \mathbb{N} \right\} \cup \left\{ \frac{p^{(1)}_{i_{2k-1}}\; p^{(o)}_{i_{2k}}}{p^{(o)}_{i_{2k-1}}\; p^{(1)}_{i_{2k}}} \; ; \; k \in \mathbb{N} \right\} .$$

Hence

$$(9) \qquad \left| \frac{d\mu^{\gamma_k}}{d\mu} \right| \leq M \quad , \quad \left| \left(\frac{d\mu^{\gamma_k}}{d\mu} \right)^{-1} \right| \leq M \quad ,$$

$$(10) \qquad \underline{\lim}_{k \to \infty} \left| \left(\frac{d\mu^{\gamma_k}}{d\mu} \right)^{-1} - 1 \right| \geqslant \varepsilon \quad .$$

These facts enable us to end the proof with an argument similar to that of L.Pukánszky ([27]).

Suppose there is a sigma-finite , Γ_ω - invariant measure ν on ω , equivalent with μ and consider

$$f = \frac{d\nu}{d\mu} \quad .$$

There exists a Borel set $F \subset \omega$ with $\mu(F) > 0$ such that f is bounded on F . Let us denote

$$f'(\alpha) = \begin{cases} f(\alpha) & \text{if} \quad \alpha \in F \cup \bigcup_{k=1}^{\infty} \gamma_k(F) \quad , \\ 0 & \text{otherwise} \quad . \end{cases}$$

Since

$$f(\gamma_k(\alpha)) \frac{d\mu^{\gamma_k}}{d\mu}(\alpha) = f(\alpha) \quad , \quad \mu\text{- almost everywhere} \; ,$$

it follows that

$$\int_\omega \left| f'(\gamma_k(\alpha)) - f'(\alpha) \right| d\mu(\alpha) \geqslant \int_F \left| f(\gamma_k(\alpha)) - f(\alpha) \right| d\mu(\alpha)$$

$$= \int_F \left| \left(\frac{d\mu^{\gamma_k}}{d\mu} \right)^{-1} - 1 \right| \left| f(\alpha) \right| d\mu(\alpha) \; .$$

By the relation (10) we infer

$$(11)\int_{\omega}\left|f'(\gamma_k(\alpha)) - f'(\alpha)\right| d\mu(\alpha) \geqslant \varepsilon \int_F |f(\alpha)| \, d\mu(\alpha) > 0 \quad .$$

On the other hand , owing to relation (9) and to the special nature of the transformations γ_k , we infer that

$$(12) \qquad \lim_{k\to\infty} \int_{\omega} \left|\varphi(\gamma_k(\alpha)) - \varphi(\alpha)\right| d\mu(\alpha) = 0$$

for any function $\varphi \in L^1(\omega, \mu)$.

The relations (11) and (12) are obviously contradictory, thus the Proposition is proved .

Q.E.D.

IV.9. Finally , consider two product measures μ and $\tilde{\mu}$ on ω defined by the sequences $\left\{p_n^{(0)}, p_n^{(1)}\right\}$ and $\left\{\tilde{p}_n^{(0)}, \tilde{p}_n^{(1)}\right\}$ respectively . Some simple necessary and sufficient conditions for the equivalence of μ and $\tilde{\mu}$ are known ([14],[15]) .

We reproduce here the result of V.Golodets ([14]) .

Let a , b be real numbers such that $0 < a < 1 < b$ and put

$$I_0 = \left\{ n \in \mathbb{N} \ ; \ a < \frac{\tilde{p}_n^{(0)}}{p_n^{(0)}} < b \right\} \quad ,$$

$$I_1 = \left\{ n \in \mathbb{N} \ ; \ a < \frac{\tilde{p}_n^{(1)}}{p_n^{(1)}} < b \right\} \quad .$$

Then <u>the measures</u> μ <u>and</u> $\tilde{\mu}$ <u>are equivalent if and only if the following conditions are simultaneously satisfied</u> :

$$\text{(i)} \qquad \sum_{n \in I_0 \cap I_1} \frac{(p_n^{(1)} - \tilde{p}_n^{(1)})^2}{p_n^{(0)} p_n^{(1)}} < +\infty \quad ,$$

$$\text{(ii)} \qquad \sum_{n \in \mathbb{N} \setminus I_0} p_n^{(0)} + \sum_{n \in \mathbb{N} \setminus I_1} p_n^{(1)} < +\infty \quad ,$$

$$\text{(iii)} \qquad \sum_{n \in \mathbb{N} \setminus I_0} \tilde{p}_n^{(0)} + \sum_{n \in \mathbb{N} \setminus I_1} \tilde{p}_n^{(1)} < +\infty \quad .$$

IV.10. By the considerations in Chapter I § 3 and Chapter II , any measure μ on ω defined as above by a sequence $\{p_n^{(0)} , p_n^{(1)}\}$ which satisfies the conditions of Propositions IV.4. and IV.8. gives rise to a factor representation of type III , of $U(\infty)$. On the other hand , if the sequence $\{p_n^{(0)} , p_n^{(1)}\}$ satisfies only the condition of Proposition IV.4. then we have an irreducible representation ρ_μ of $U(\infty)$. In what follows <u>we shall construct concrete realisations of these types of representations</u> .

Thus , fix a sequence $\{p_n^{(0)} , p_n^{(1)}\}$ such that

$$0 < p_n^{(0)} < 1 \quad , \quad 0 < p_n^{(1)} < 1 \quad , \quad p_n^{(0)} + p_n^{(1)} = 1$$

and put

$$q_n = \frac{p_n^{(1)}}{p_n^{(0)}} \quad .$$

Denote by μ the corresponding product measure on ω .

IV.11. Consider a separable Hilbert space H with an

orthonormal basis $\{e_n\}$. We define

$$H_n = \bigoplus_{j=1}^{n} \mathbb{C}\, e_j \ .$$

We shall identify $GL(n,\mathbb{C})$ with the set of invertible elements $T \in L(H)$ such that

$$T e_j = e_j \quad \text{for } j > n \ .$$

Thus , $GL(\infty,\mathbb{C})$ and $U(\infty)$ can be identified with subgroups of $GL(H)$ and $U(H)$, respectively .

We shall use the following bounded linear operators on H :

$$T_n \in GL(n,\mathbb{C}) \quad ; \quad T_n e_j = \sqrt{q_j}\, e_j \ , \ j = 1,2,\dots,n \ ;$$

$$A_n \in GL(n,\mathbb{C}) \quad ; \quad A_n e_j = p_j^{(1)} e_j \ , \ j = 1,2,\dots,n \ ;$$

$$A \in GL(\infty,\mathbb{C}) \quad ; \quad A e_j = p_j^{(1)} e_j \ , \ j \in \mathbb{N} \ .$$

It is clear that

$$(I - A_n)\, T_n^{\,2} = A_n \ ,$$

$$(I - A) + A V_n = (I - A_n) + A_n V_n \ , \text{ for all } V_n \in L(H_n) \ .$$

IV.12. We denote by $\overset{k}{\bigwedge} H_n$ the kth exterior power of H_n and we consider the Hilbert spaces

$$Y_n = \bigoplus_{k=0}^{n} \overset{k}{\bigwedge} H_n \ ,$$

$$X_n = \bigoplus_{k=0}^{n} \overset{k}{\bigwedge} H_n \otimes \overset{k}{\bigwedge} H_n \ .$$

On Y_n we consider also the exterior multiplication , denoted by " $\wedge$ " ,and on X_n we consider a multiplication , denoted again by " $\wedge$ " ,but given by the rule

$$(a_i \otimes b_i) \wedge (a_j \otimes b_j) = (a_i \wedge a_j) \otimes (b_i \wedge b_j) \quad ,$$

$$a_i \, , \, b_i \in \bigwedge^{i} H_n \quad , \quad a_j \, , \, b_j \in \bigwedge^{j} H_n \quad .$$

With these conventions we define the vectors

$$\theta_n \, , \, \eta_n \in Y_n \quad , \quad \zeta_n \, , \, \xi_n \in X_n$$

as follows :

$$\theta_n = \bigwedge_{j=1}^{n} (1 + e_j) = \sum_{k=0}^{n} \sum_{i_1 < \ldots < i_k} e_{i_1} \wedge \ldots \wedge e_{i_k}$$

$$\eta_n = \bigwedge_{j=1}^{n} \left(\sqrt{p^{(o)}} . 1 + \sqrt{p^{(1)}} . e_j \right) =$$

$$= \sqrt{p_1^{(o)} \ldots p_n^{(o)}} \sum_{k=0}^{n} \sum_{i_1 < \ldots < i_k} \sqrt{q_{i_1} \ldots q_{i_k}} \; e_{i_1} \wedge \ldots \wedge e_{i_k}$$

$$\zeta_n = \bigwedge_{j=1}^{n} (1 \otimes 1 + e_j \otimes e_j) = \sum_{k=0}^{n} \sum_{i_1 < \ldots < i_k} (e_{i_1} \wedge \ldots \wedge e_{i_k}) \otimes (e_{i_1} \wedge \ldots \wedge e_{i_k})$$

$$\xi_n = \bigwedge_{j=1}^{n} \left(\sqrt{p_j^{(o)}} \, 1 \otimes 1 + \sqrt{p_j^{(1)}} \, e_j \otimes e_j \right) =$$

$$= \sqrt{p_1^{(o)} \ldots p_n^{(o)}} \sum_{k=0}^{n} \sum_{i_1 < \ldots < i_k} \sqrt{q_{i_1} \ldots q_{i_k}} \; (e_{i_1} \wedge \ldots \wedge e_{i_k}) \otimes (e_{i_1} \wedge \ldots \wedge e_{i_k})$$

IV.13. There is a natural irreducible representation δ_n^k of $GL(n,\mathbb{C})$ on $\bigwedge^{k} H_n$, namely that with signature

$$(\underbrace{1, \ldots, 1}_{k\text{-times}}, \underbrace{0, \ldots \ldots, 0}_{(n-k)\text{-times}}) \quad .$$

Hence we have the representations

$$\delta_n = \bigoplus_{k=0}^{n} \delta_n^k \qquad \text{of} \quad GL(n,\mathbb{C}) \quad \text{on} \quad Y_n \quad ,$$

$$\sigma_n = \bigoplus_{k=o}^{n} \delta_n^k \otimes I_n^k \quad \text{of} \quad GL(n,\mathbb{C}) \quad \text{on} \quad X_n ,$$

where I_n^k stands for the trivial representation of $GL(n,C)$ on $\bigwedge^k H_n$ We shall denote by the same symbol the restriction of a representation from $GL(n,\mathbb{C})$ to $U(n)$ and its extension to the measure algebra.

It is easy to verify that θ_n is a cyclic vector for the representation δ_n and that ζ_n is a cyclic vector for the representation σ_n . Since

$$\delta_n(T_n)\theta_n = \sum_{k=o}^{n} \sum_{i_1<\ldots<i_k} T_n e_{i_1}\wedge\ldots\wedge T_n e_{i_k} = (\sqrt{p_1^{(o)}\ldots p_n^{(o)}})^{-1}\eta_n ,$$

$$\sigma_n(T_n)\zeta_n = \sum_{k=o}^{n} \sum_{i_1<\ldots<i_k} (T_n e_{i_1}\wedge\ldots\wedge T_n e_{i_k}) \otimes (e_{i_1}\wedge\ldots\wedge e_{i_k}) =$$

$$= (\sqrt{p_1^{(o)}\ldots p_n^{(o)}})^{-1}\xi_n ,$$

it follows that

η_n is a cyclic vector for the representation δ_n ,

ξ_n is a cyclic vector for the representation σ_n .

IV.14. By the way we have obtained

$$\delta_n(T_n^{-1})\eta_n = \det(I - A_n)^{1/2}\,\theta_n ,$$

$$\sigma_n(T_n^{-1})\xi_n = \det(I - A_n)^{1/2}\,\zeta_n .$$

For any $V \in GL(n,\mathbb{C})$ we have

$$(\sigma_n(V)\zeta_n \mid \zeta_n) =$$

$$= \left(\sum_{k=0}^{n} \sum_{i_1 < \ldots < i_k} (Ve_{i_1} \wedge \ldots \wedge Ve_{i_k}) \otimes (e_{i_1} \wedge \ldots \wedge e_{i_k}) \right|$$

$$\left| \sum_{k=0}^{n} \sum_{i_1 < \ldots < i_k} (e_{i_1} \wedge \ldots \wedge e_{i_k}) \otimes (e_{i_1} \wedge \ldots \wedge e_{i_k}) \right)$$

$$= \sum_{k=0}^{n} \sum_{i_1 < \ldots < i_k} \left(Ve_{i_1} \wedge \ldots \wedge Ve_{i_k} \middle| e_{i_1} \wedge \ldots \wedge e_{i_k} \right)$$

$$= \mathrm{Tr}_{Y_n} (\sigma_n(V)) \quad ,$$

where Tr_{Y_n} denotes the natural trace on $L(Y_n)$. This shows that

$$V \longmapsto (\sigma_n(V) \zeta_n | \zeta_n)$$

is a central function on $GL(n,\mathbb{C})$ and that if V is diagonal ,

$$V = \begin{pmatrix} \lambda_1 & & \\ & \ddots & \\ & & \lambda_n \end{pmatrix} \quad ,$$

we have

$$(\sigma_n(V) \zeta_n | \zeta_n) = \sum_{k=0}^{n} \sum_{i_1 < \ldots < i_k} \lambda_{i_1} \cdots \lambda_{i_k}$$

$$= \prod_{i=1}^{n} (1 + \lambda_i)$$

$$= \det (I + V) \quad .$$

Since

$$V \longmapsto \det (I + V)$$

is also a central function on $GL(n,\mathbb{C})$, it follows that

$$(\sigma_n(V) \zeta_n | \zeta_n) = \det (I + V) \quad \text{for all} \quad V \in GL(n,\mathbb{C}) \; .$$

Therefore , for all $V \in GL(n,\mathbb{C})$ we have

$$(\sigma_n(V)\xi_n \mid \xi_n) = \det(I - A_n)\ (\sigma_n(T_n V\, T_n)\zeta_n \mid \zeta_n)$$

$$= \det(I - A_n)\ \det(I + T_n V\, T_n)$$

(13) $$= \det(I - A_n)\ \det(I + T_n^{\,2}\, V)$$

$$= \det((I - A_n) + A_n V)$$

$$= \det((I - A) + A\,V) \qquad .$$

IV.15. Since $H_n \subset H_{n+1}$, we may consider

$$Y_n \subset Y_{n+1} \quad , \quad X_n \subset X_{n+1} \qquad .$$

The isometric linear maps

$$J_n : Y_n \longrightarrow Y_{n+1} \quad , \quad I_n : X_n \longrightarrow X_{n+1}$$

defined by

$$J_n(\eta) = \eta \wedge (\sqrt{p_{n+1}^{(o)}} \cdot 1 + \sqrt{p_{n+1}^{(1)}} \cdot e_{n+1}) \quad , \ \eta \in Y_n \ ,$$

$$I_n(\xi) = \xi \wedge (\sqrt{p_{n+1}^{(o)}} \cdot 1 \otimes 1 + \sqrt{p_{n+1}^{(1)}} \cdot e_{n+1} \otimes e_{n+1}) \ , \ \xi \in X_n \ ,$$

allow us to consider

Y = the Hilbert space direct limit of the Y_n's following the J_n's ,

X = the Hilbert space direct limit of the X_n's following the I_n's .

Since $J_n \eta_n = \eta_{n+1}$ and $I_n \xi_n = \xi_{n+1}$, we may define

$$\eta_o = \varinjlim \eta_n \in Y \quad , \quad \xi_o = \varinjlim \xi_n \in X \qquad .$$

For all $U \in U(n) \subset U(n+1)$ we have

$$\delta_{n+1}(U) \circ J_n = J_n \circ \delta_n(U) \quad , \quad \sigma_{n+1}(U) \circ I_n = I_n \circ \sigma_n(U).$$

Therefore we get the following representations of $U(\infty)$ on Y and X , respectively :

δ = the direct limit of the representations δ_n ,

σ = the direct limit of the representations σ_n .

<u>We shall denote by the same symbol the corresponding representations</u>

of $A(U(\infty))$. We remark that

η_0 is a cyclic vector for the representation δ ,

ξ_0 is a cyclic vector for the representation σ .

IV.16. Our aim is to show that σ is unitarily equivalent to π_μ and that δ is unitarily equivalent to ρ_μ corresponding to a suitable system of matrix units $E_{\alpha,\beta}$. Only the first verification will be done in full detail .

Thus , for the equivalence of the representations σ and π_μ , using the results obtained in Section II.2.10. , we have to verify that :

$$(14) \qquad (\sigma_n(\delta_g * p_{\rho_n})\xi_n \; \xi_n) = \sum_{\alpha \in S(\rho_n)} \mu(p_\alpha) \, \mathrm{Tr}_{B_{\rho_n}}(p_\alpha * \delta_g * p_\alpha)$$

for all $g \in U(n)$, $\rho_n \in \widehat{U(n)}$ and $n \in \mathbb{N}$.

If ρ_n does not correspond to a signature of the form

$$(1,\ldots,1,0,\ldots\ldots,0)$$

then we have

$$\sigma_n(p_{\rho_n}) = 0 \ , \ \delta_n(p_{\rho_n}) = 0 \ \text{ and } \ \mu(p_{\rho_n}) = 0 \ ,$$

so the above equality is trivially satisfied .

Suppose now that ρ_n corresponds to the signature

$$(\underbrace{1,\ldots,1}_{k_n\text{-times}},\underbrace{0,\ldots\ldots,0}_{(n-k_n)\text{-times}}) \quad , \quad 0 \leq k_n \leq n \quad .$$

Then ρ_n is (equivalent to) the natural representation of $U(n)$ on $\bigwedge^{k_n} H_n$, hence ρ_n defines a $*$ - isomorphism

$$\rho_n : B_{\rho_n} \longrightarrow L(\bigwedge^{k_n} H_n) \quad .$$

Consider

$$\alpha = (\rho_1 \prec \rho_2 \prec \dots \prec \rho_{n-1} \prec \rho_n) \in S(\rho_n) \quad ,$$

where

$$\rho_j = (\underbrace{1,\dots,1}_{k_j\text{-times}},\underbrace{0,\dots\dots,0}_{(j-k_j)\text{-times}}) \quad , \quad 1 \leq j \leq n \quad .$$

Define $(\alpha_j)_{j=1}^n \in \{0,1\}^n$ such that

$$\sum_{s=1}^{j} \alpha_s = k_j \qquad , \quad 1 \leq j \leq n \quad ,$$

and denote by

$$1 \leq i_1 < i_2 < \dots < i_{k_n} \leq n$$

those indices for which

$$\alpha_{i_m} = 1 \quad , \quad m = 1,2,\dots,k_n \quad .$$

The reader should keep in mind the correspondences

$$\alpha \longleftrightarrow \{k_1,\dots,k_n\} \longleftrightarrow (\alpha_j)_{j=1}^n \longleftrightarrow \{i_1 < i_2 < \dots < i_{k_n}\} \quad .$$

We denote

$$e_\alpha = e_{i_1} \wedge e_{i_2} \wedge \dots \wedge e_{i_{k_n}} \in \bigwedge^{k_n} H_n \quad .$$

It is easy to verify that

$$\rho_n(p_\alpha) = \text{the orthogonal projection of } \bigwedge^{k_n} H_n \text{ onto } \mathbb{C}\, e_\alpha \, .$$

Since $\{e_\alpha\}_{\alpha \in S(\rho_n)}$ is an orthonormal basis in $\bigwedge^{k_n} H_n$, we have

$$\mathrm{Tr}_{B_{\rho_n}}(x) = \mathrm{Tr}\, \rho_n(x) = \sum_{\alpha \in S(\rho_n)} (\rho_n(x)\, e_\alpha \mid e_\alpha) \quad , \quad x \in B_{\rho_n} \quad .$$

On the other hand, owing to the identification $\omega = \{0,1\}^{\aleph_0}$ we find

$$\mu(p_\alpha) = \prod_{j=1}^{n} p_j^{(\alpha_j)} = \left(\prod_{j=1}^{n} p_j^{(0)}\right)\left(\prod_{s=1}^{k_n} q_{i_s}\right) .$$

We proceed to verify (14). By the above we have

$$\sum_{\alpha \in S(\rho_n)} \mu(p_\alpha)\, \mathrm{Tr}_{B_{\rho_n}} (p_\alpha * \delta_g * p_\alpha) =$$

$$= \left(\prod_{j=1}^{n} p_j^{(0)}\right) \sum_{\alpha \in S(\rho_n)} \left(\prod_{s=1}^{k_n} q_{i_s}\right) (\rho_n(g)\, e_\alpha \mid e_\alpha).$$

Since $\sigma_n(p_{\rho_n})$ is the orthogonal projection of X_n onto the isotypic component of type ρ_n of σ_n, we have

$$\sigma_n(p_{\rho_n})\xi_n = \left(\prod_{j=1}^{n} \sqrt{p_j^{(0)}}\right) \sum_{i_1<\dots<i_{k_n}} \left(\prod_{s=1}^{k_n} \sqrt{q_{i_s}}\right) (e_{i_1} \wedge \dots \wedge e_{i_{k_n}}) \otimes (e_{i_1} \wedge \dots \wedge e_{i_{k_n}})$$

hence

$$(\sigma_n(\delta_g * p_{\rho_n})\, \xi_n \mid \xi_n) = (\sigma_n(g)\, \sigma_n(p_{\rho_n})\, \xi_n \mid \sigma_n(p_{\rho_n})\, \xi_n) =$$

$$= \left(\prod_{j=1}^{n} p_j^{(0)}\right) \left(\sum_{\alpha \in S(\rho_n)} \left(\prod_{s=1}^{k_n} \sqrt{q_{i_s}}\right) \rho_n(g)\, e_\alpha \otimes e_\alpha \,\middle|\, \sum_{\alpha \in S(\rho_n)} \left(\prod_{s=1}^{k_n} \sqrt{q_{i_s}}\right) e_\alpha \otimes e_\alpha \right)$$

$$= \left(\prod_{j=1}^{n} p_j^{(0)}\right) \sum_{\alpha \in S(\rho_n)} \left(\prod_{s=1}^{k_n} q_{i_s}\right) (\rho_n(g)\, e_\alpha \mid e_\alpha) .$$

Therefore <u>the representations σ and π_μ are unitarily equivalent</u>.

Let us also indicate how for δ and ρ_μ we can proceed similarly . First we specify the system of matrix units $E_{\alpha,\beta}$ by means of which we identify A and $A(\Omega,\Gamma)$. It will be sufficient to do this only for $\alpha,\beta \in S(\rho_n)$, $n \in \mathbb{N}$, where ρ_n is of the form

$$(\underbrace{1,\ldots,1}_{k_n\text{-times}},\underbrace{0,\ldots\ldots,0}_{(n-k_n)\text{-times}}) \qquad , \quad 0 \leqslant k_n \leqslant n < +\infty .$$

We choose the $E_{\alpha,\beta}$'s such that

$$\rho_n(E_{\alpha,\beta})\, e_\beta = e_\alpha \quad ; \ \alpha,\beta \in S(\rho_n) \quad , \quad 0 \leqslant k_n \leqslant n < +\infty .$$

Then , restricting again the verifications to the B_{ρ_n} 's , one can prove that δ <u>and</u> ρ_μ <u>are unitarily equivalent</u> by showing that <u>the states determined by</u> η <u>and respectively by the vector</u>

$$1 \in L^2(\omega,\mu)$$

<u>are equal</u> .

IV.17. Taking into account Section I.3.16. , Proposition IV.4. and the last result in Section IV.16. , we derive a first conclusion :

THEOREM . <u>The representation</u> δ <u>of</u> $U(\infty)$ <u>on the Hilbert space</u> Y <u>constructed as above is irreducible provided that</u> :

$$\sum_{n=1}^{\infty} \min\left\{p_n^{(0)} , p_n^{(1)}\right\} = +\infty .$$

IV.18. Concerning the representation σ , we first remark that the function of positive type on $U(\infty)$ associated to σ and ξ has a simple expression . Indeed , from relation (13) we infer

$$(\sigma(U)\xi \mid \xi) = \det((I - A) + AU) \quad ; \quad U \in U(\infty) .$$

Thus , the function

$$\Psi_A(U) = \det((I - A) + AU) \quad ; \quad U \in U(\infty)$$

is of positive type on $U(\infty)$ and , denoting by π_A the representation of $U(\infty)$ associated to Ψ_A , we have

$$\pi_A \simeq \sigma \simeq \pi_\mu .$$

Recall that here $A \in GL(\infty,\mathbb{C}) \subset L(H)$ is defined by

$$A e_j = a_j e_j \quad ; \quad 0 < a_j < 1 \quad , \quad j \in \mathbb{N} \quad ,$$

where $a_j = p_j^{(1)}$, $j \in \mathbb{N}$.

The function $T \longmapsto \det T$ is defined for all operators $T \in L(H)$ of the form $T = I + N$, with N nuclear operator . It follows that the function Ψ_A is defined on $U_1(\infty)$.Moreover , it is easy to see that Ψ_A is continuous on $U_1(\infty)$ with respect to the topology defined by the metric

$$d(U',U'') = \operatorname{Tr} |U' - U''| \quad , \quad U',U'' \in U_1(\infty) \quad .$$

Since $U(\infty)$ is dense in $U_1(\infty)$, Ψ_A is a function of positive type on $U_1(\infty)$.

Thus , the representation π_A extends to a representation of $U_1(\infty)$, denoted also by π_A , which is associated to Ψ_A . Using again the density of $U(\infty)$ in $U_1(\infty)$ we obtain that the von Neumann algebras generated by $\pi_A(U(\infty))$ and by $\pi_A(U_1(\infty))$ are equal .

Let V be a unitary operator on H . Then

$$i_V : U_1(\infty) \ni U \longmapsto V^*UV \in U_1(\infty)$$

is an automorphism of $U_1(\infty)$ and

$$(\psi_A \circ i_V)(U) = \psi_{V^*AV}(U) \quad ; \quad U \in U_1(\infty) \ .$$

Therefore ψ_{V^*AV} is also a function of positive type on $U_1(\infty)$ and its associated representation ψ_{V^*AV} coincides with $\psi_A \circ i_V$.

Since the von Neumann algebras generated by $\pi_{V^*AV}(U_1(\infty))$ and by $\pi_A(U_1(\infty))$ are the same, it follows that the representations π_{V^*AV} and π_A are both of the same type (without being necessarily equivalent).

Thus, for any injective operator $A \in L(H)$, $0 \leq A \leq I$, diagonable with respect to <u>some</u> orthonormal basis of H, the function ψ_A is of positive type. By the spectral theorem one can easily check that any operator $A \in L(H)$, $0 \leq A \leq I$, is the norm limit of a sequence of diagonable injective operators $A_n \in L(H)$, $0 \leq A_n \leq I$. It follows that

PROPOSITION. *For any operator* $A \in L(H)$, $0 \leq A \leq I$, *the function*

$$\psi_A : U_1(\infty) \ni U \longmapsto \det\left((I - A) + AU\right) \in \mathbb{C}$$

is continuous and of positive type on $U_1(\infty)$.

Denote again by π_A the representation of $U_1(\infty)$ associated to ψ_A. The following problem naturally arises:

PROBLEM. *Under which conditions on* A *is the representation* π_A *factorial? Which is the type of the corresponding factor? When are two such representations equivalent?*

As we remarked above, the representations π_A and π_{V^*AV}, $V \in U(H)$, are of the same type. Therefore, *the type of* π_A

depends only on the spectral properties of A . However , it seems difficult to decide when π_A and π_{V^*AV} are equivalent .

IV.19. The considerations concerning the representation σ allow us to give a partial answer to the above problem. Indeed , taking into account Theorem I.3.12., Proposition IV.4., Proposition IV.8. and the first conclusion in Section IV.16., we obtain the following

THEOREM . *The representation* σ *of* $U(\infty)$ *on the Hilbert space* X *constructed as above is factorial provided that*

$$\sum_{n=1}^{\infty} \min \{p_n^{(0)} , p_n^{(1)}\} = +\infty$$

and the corresponding factor is of type III *provided that*

the sequence $\{p_n^{(0)}\}$ *has two different limit points in* $(0,1)$.

Note that necessary and sufficient conditions for the equivalence of two such representations are contained in Section IV.9.

COROLLARY . *Let* $A \in L(H)$ *be a diagonable operator with eigenvalues* $\{a_j\}_{j\in \mathbb{N}}$ *such that*

$$0 < a_j < 1 \quad , \quad j \in \mathbb{N} .$$

Then the representation π_A *of* $U(\infty)$ *associated to the function of positive type*

$$\psi_A(U) = \det ((I - A) + AU) \quad , \quad U \in U(\infty) ,$$

is factorial provided that

$$\sum_{j=1}^{\infty} \min \{a_j , 1 - a_j\} = + \infty$$

<u>and the corresponding factor is of type</u> III <u>provided that</u>

<u>the sequence</u> $\{a_j\}$ <u>has two different limit points in</u> (0,1) .

The last condition in the Corollary means that the essential spectrum of A has (at least) two different points in (0,1) .

IV.20. If $p_j^{(1)} = a_j = a = \text{constant}$, then the corresponding product measure μ on ω is Γ_ω - invariant , so the representations $\pi_A \simeq \sigma \simeq \pi_\mu$ are of type II_1 and they correspond to the character

$$U \longmapsto \det ((1 - a) + aU) \quad , \quad U \in U(\infty).$$

Type II_1 -factor representations of $U(\infty)$ were studied in ([34],[35]) , where also other classes of such representations were obtained . It would be interesting for instance to extend also the construction of type II_1 - factor representations in symmetric tensors , as was done here for antisymmetric ones .

IV.21. Finally , let us mention that <u>the modular automorphism group</u> $\{\sigma_t\}_{t\in\mathbb{R}}$ <u>of the von Neumann algebra</u> $(\pi_A(U(\infty)))'' = (\sigma(U(\infty)))''$ ($Ae_j = a_j e_j$, $0 < a_j < 1$) <u>corresponding to the cyclic separating vector</u> ξ_0 (see [32]) <u>has a group interpretation</u>

Namely , with $B = A(I - A)^{-1}$, consider the following automorphisms of $U(\infty)$:

$$s_t(V) = B^{it} V B^{-it} \quad , \quad V \in U(\infty) \quad ; \quad t \in \mathbb{R} \quad .$$

Then :

$$(15) \qquad \sigma_t(\pi_A(V)) = \pi_A(s_t(V)) \quad , \quad V \in U(\infty) \quad ; \quad t \in \mathbb{R} \quad .$$

This can be verified by reduction to the case of

$$((\pi_A \mid U(n)) \mid X_n)'' \subset L(X_n) \quad \text{and} \quad \xi_n \in X_n \quad .$$

The reduction is possible since X_n is invariant for the closure of the operator $(\pi_A(U(\infty)))''\xi_o \quad x\xi_o \longmapsto x^*\xi_o$ and for its adjoint , which is the analogous operator with respect to the commutant .

CHAPTER V SOME TYPE II_∞ FACTOR REPRESENTATIONS OF $U(\infty)$

§ 1 Infinite tensor product representations

In this section we shall show that some natural representations of $U(\infty)$ on the infinite tensor products of the underlying Hilbert space are factor representations of type II_∞ and we shall study the equivalence of different such representations as well as their commutation properties with the corresponding natural representations of the group $S(\infty)$ of finite permutations of $\mathbb{N}$.

This section does not depend on the rest of the work .

V.1.1. Consider a separable Hilbert space H with an orthonormal basis $\{e_n\}_{n \in \mathbb{N}}$, define

$$H_n = \bigoplus_{j=1}^{n} \mathbb{C}\, e_j \quad , \quad n \in \mathbb{N} \quad ,$$

and , as usually , identify $GL(n,\mathbb{C})$ with the set of invertible operators $T \in L(H)$ acting identically on $H \ominus H_n$. Then $GL(\infty,\mathbb{C})$ and $U(\infty)$ can be identified with subgroups of $GL(H)$ and $U(H)$, respectively .

Consider also the Hilbert spaces :

$$\mathcal{H}_n^m = \bigotimes^{m} H_n = \underbrace{H_n \otimes \cdots \otimes H_n}_{m\text{-times}} \quad ; n,m \in \mathbb{N}$$

There are natural representations μ_n^m and ν_n^m of the group $U(n)$ and respectively of the symmetric group $S(m)$ on the Hilbert space $\mathcal{H}_n^m$ such that

$$\mu_n^m(U)\left(\bigotimes_{j=1}^{m} \xi_j\right) = \bigotimes_{j=1}^{m} U\xi_j \quad ; \xi_1, \dots, \xi_m \in H_n \quad , \quad U \in U(n) \quad ,$$

$$\nu_n^m(\sigma)\left(\bigotimes_{j=1}^{m} \xi_j\right) = \bigotimes_{j=1}^{m} \xi_{\sigma^{-1}(j)} \quad ; \xi_1, \dots, \xi_m \in H_n \quad , \quad \sigma \in S(m) \quad .$$

Denote by

$$M_n^m \quad \text{and} \quad N_n^m$$

the von Neumann algebras generated in $L(\mathcal{H}_n^m)$ by

$$\mu_n^m(U(n)) \quad \text{and} \quad \nu_n^m(S(m))$$

respectively . Let us recall the following classical result ([36])

THEOREM (Hermann Weyl) . <u>The von Neumann algebra</u> N_n^m <u>is the commutant of the von Neumann algebra</u> M_n^m <u>in</u> $L(\mathcal{H}_n^m)$:

$$(M_n^m)' = N_n^m \quad .$$

If $\xi_1 , \dots , \xi_m \in H_n$ are linear independent vectors , then the vectors

$$\nu_n^m(\sigma)\left(\bigotimes_{j=1}^{m} \xi_j\right) = \bigotimes_{j=1}^{m} \xi_{\sigma^{-1}(j)} \quad , \quad \sigma \in S(m) \quad ,$$

are linear independent in $\mathcal{H}_n^m$. Since N_n^m is linearly spaned by $\left\{\nu_n^m(\sigma) \; ; \sigma \in S(m)\right\}$, it follows that

$$\bigotimes_{j=1}^{m} \xi_j \quad \text{is a separating vector for} \quad N_n^m \quad ,$$

that is

$$y \in N_n^m \quad , \quad y\left(\bigotimes_{j=1}^{m} \xi_j\right) = 0 \Longrightarrow y = 0 \quad .$$

But a vector is separating for a von Neumann algebra if and only if it is cyclic for the commutant . Thus , by the above theorem we see that

$$\bigotimes_{j=1}^{m} \xi_j \quad \text{is a cyclic vector for} \quad M_n^m \quad .$$

V.1.2. We fix a strictly increasing sequence κ of positive integers

$$\kappa = (1 \leq k_1 < k_2 < \dots < k_n < k_{n+1} < \dots) \quad ,$$

we define the Hilbert spaces

$$H^{(n)} = \bigotimes^{n} H = \underbrace{H \otimes \dots \otimes H}_{n\text{-times}} \quad , \quad n \in \mathbb{N} \quad ,$$

and the linear isometric maps

$$I_n^\kappa : H^{(n)} \ni \xi \longmapsto \xi \otimes e_{k_{n+1}} \in H^{(n+1)} \quad , \quad n \in \mathbb{N} .$$

We consider the Hilbert space direct limit

$$\mathcal{H}^\kappa$$

of the $H^{(n)}$'s following the I_n 's .

The Hilbert space $\mathcal{H}^\kappa$ is nothing but von Neumann's infinite tensor product of a sequence of copies of H along the sequence of vectors

$$e_{k_1} , e_{k_2} , \dots , e_{k_n} , \dots \quad \in \quad H$$

Let us recall that for any sequence of vectors

$$\xi_1 , \xi_2 , \dots , \xi_n , \dots \quad \in \quad H$$

such that

$$\sum_{j=1} | 1 - \|\xi_j\| | < +\infty \quad \text{and} \quad \sum_{j=1} | 1 - (\xi_j | e_{k_j}) | < +\infty$$

there corresponds a "decomposable vector"

$$\bigotimes_{j=1}^{\infty} \xi_j = \lim_{n \to \infty} \xi_1 \otimes \dots \otimes \xi_n \otimes e_{k_{n+1}} \otimes e_{k_{n+2}} \otimes \dots \in \mathcal{H}^\kappa$$

which depends linearly on each ξ_j . If $\bigotimes_{j=1}^{\infty} \eta_j$ is another decomposable vector , then

$$\left(\bigotimes_{j=1}^{\infty} \xi_j \,\middle|\, \bigotimes_{j=1}^{\infty} \eta_j\right) = \prod_{j=1}^{\infty} (\xi_j | \eta_j)$$

where the infinite product is absolutely convergent . The set of decomposable vectors is total in $\mathcal{H}^K$.

Let $S(\infty)$ be the discrete group of finite permutations of $\mathbb{N}$, that is, the obvious direct limit of the symmetric groups $S(m)$. There are natural unitary representations

$$\mu^K \quad \text{and} \quad \nu^K$$

of the groups $U(\infty)$ and respectively $S(\infty)$ on $\mathcal{H}^K$ such that

$$\mu^K(U)\left(\bigotimes_{j=1}^{\infty} \xi_j\right) = \bigotimes_{j=1}^{\infty} U\xi_j \qquad , \quad U \in U(\infty) \quad ,$$

$$\nu^K(\sigma)\left(\bigotimes_{j=1}^{\infty} \xi_j\right) = \bigotimes_{j=1}^{\infty} \xi_{\sigma^{-1}(j)} \qquad , \quad \sigma \in S(\infty) \quad ,$$

for any decomposable vector

$$\bigotimes_{j=1}^{\infty} \xi_j \in \mathcal{H}^K \quad .$$

V.1.3. Before going any further we shall prove a simple lemma on the commutants of von Neumann algebras . For a von Neumann algebra $M \subset L(\mathcal{H})$ we denote by M' its commutant . Also , if $p \in M'$ (resp. $p \in M$) is a projection , then we denote by $M_p \subset L(p\mathcal{H})$ the induced (resp. the reduced) von Neumann algebra. We recall that

$$(M')_p = (M_p)' \quad .$$

LEMMA . <u>Let</u> $M , N \subset L(\mathcal{H})$ <u>be von Neumann algebras such</u>

that $N \subset M'$. *Suppose there are*

(i) *an increasing sequence* $\{M_n\}$ *of von Neumann subalgebras of* M *which generates* M , *i.e.* $M = (\bigcup_{n=1}^{\infty} M_n)''$;

(ii) *an increasing sequence* $\{N_n\}$ *of von Neumann subalgebras of* N *which generates* N , *i.e.* $N = (\bigcup_{n=1}^{\infty} N_n)''$;

(iii) *an increasing sequence of projections* $p_n \in M_n' \cap N_n'$;

such that

$$p_n \uparrow 1 \quad \text{and} \quad ((M_n)_{p_n})' = (N_n)_{p_n} \quad \text{for each } n .$$

Then

$$M' = N .$$

Proof . We have to prove that $M' \subset N'' = N$. Thus , consider $x \in M'$ and $y \in N'$. For each n we have

$$x \in M_n' \quad , \quad y \in N_n'$$

so , by our assumptions ,

$$p_n x p_n \in ((M_n)_{p_n})' = (N_n)_{p_n} \quad , \quad p_n y p_n \in ((N_n)_{p_n})' .$$

It follows that

$$p_n x p_n y p_n = p_n y p_n x p_n \quad \text{for each } n .$$

Taking the limit when $n \longrightarrow \infty$, we get

$$xy = yx .$$

Since $y \in N'$ was arbitrary , we obtain $x \in N''$.

Q.E.D.

V.1.4. We denote by

$$M^{\kappa} \quad \text{and} \quad N^{\kappa}$$

the von Neumann algebras generated in $L(\mathcal{H}^{\kappa})$ by

$$\mu^{\kappa}(U(\infty)) \text{ and } \nu^{\kappa}(S(\infty))$$

respectively. We shall prove

THEOREM . _The von Neumann algebra_ N^κ _is the commutant of the von Neumann algebra_ M^κ _in_ $L(\mathcal{H}^\kappa)$:

$$(M^\kappa)' = N^\kappa .$$

Proof . Clearly , $(M^\kappa)' \supset N^\kappa$. For each n denote by

$$M_n \quad \text{and} \quad N_n$$

the von Neumann algebras generated in $L(\mathcal{H}^\kappa)$ by

$$\mu^\kappa(U(k_n)) \quad \text{and} \quad \nu^\kappa(S(n))$$

respectively . Then , obviously , M_n and N_n fulfil the conditions (i) and (ii) in Lemma V.1.3.

On the other hand we have

$$\mathcal{H}^n_{k_n} \subset H^{(n)} \subset \mathcal{H}^\kappa , \quad \mathcal{H}^n_{k_n} \subset \mathcal{H}^{n+1}_{k_{n+1}} , \quad \overline{\bigcup_{n=1}^{\infty} \mathcal{H}^n_{k_n}} = \mathcal{H}^\kappa$$

and $\mathcal{H}^n_{k_n}$ is an invariant subspace for both the representations $\mu^\kappa | U(k_n)$ and $\nu^\kappa | S(n)$. Moreover , it is clear that the representation $\mu^\kappa | U(k_n)$ induces on $\mathcal{H}^n_{k_n}$ the representation $\mu^n_{k_n}$ and the representation $\nu^\kappa | S(n)$ induces on $\mathcal{H}^n_{k_n}$ the representation $\nu^n_{k_n}$.

Denote by p_n the orthogonal projection of $\mathcal{H}^\kappa$ onto $\mathcal{H}^n_{k_n}$. Then we have

$$p_n \uparrow 1 , \quad p_n \in M_n' \cap N_n' ,$$

and

$$(M_n)_{p_n} = M^n_{k_n} , \quad (N_n)_{p_n} = N^n_{k_n} ,$$

thus , by Hermann Weyl ' s theorem ,

$$((M_n)_{p_n})' = (N_n)_{p_n} .$$

Therefore , all the assumptions in Lemma V.1.3. are fulfiled and the Theorem follows .

Q.E.D.

V.1.5. THEOREM . The representation μ^{κ} of $U(\infty)$ on $\mathcal{H}^{\kappa}$ is a factor representation of type II_{∞} .

Proof . We shall show that N^{κ} is a type II_1 factor and that M^{κ} is a type II_{∞} factor .

By the remark at the end of Section V.1.1. it follows that $\bigotimes_{j=1}^{n} e_{k_j} \in \mathcal{H}^{n}_{k_n}$ is a cyclic vector for $M^{n}_{k_n}$, $n \in \mathbb{N}$. Therefore

$$\xi_0^{\kappa} = \bigotimes_{j=1}^{\infty} e_{k_j} \in \mathcal{H}^{\kappa}$$

is a cyclic vector for M^{κ} . Since $N^{\kappa} = (M^{\kappa})'$, ξ_0^{κ} is a separating vector for N^{κ} .

On the other hand , for any $\sigma \in S(\infty)$ we have

$$(\nu^{\kappa}(\sigma)\,\xi_0^{\kappa} \mid \xi_0^{\kappa}) = \begin{cases} 1 & \text{if } \sigma = \varepsilon \\ 0 & \text{if } \sigma \neq \varepsilon \end{cases}$$

where $\varepsilon \in S(\infty)$ is the identity permutation . It follows that the vector state

$$\varphi^{\kappa} = \omega_{\xi_0^{\kappa}}$$

is a faithful finite normal trace on the von Neumann algebra N^{κ} .

Let us recall that the von Neumann algebra generated by the left regular representation of $S(\infty)$ is the hyperfinite type II_1 factor and that the canonical trace on this factor induces the same function of positive type on $S(\infty)$ as the above φ^{κ} does (see [28]) . It follows that the representation of $S(\infty)$

induced by ν^κ on the invariant subspace $[N^\kappa \xi_0^\kappa]$ of $\mathcal{H}^\kappa$ is unitarily equivalent to the left regular representation of $S(\infty)$.

Denote by p the orthogonal projection of $\mathcal{H}^\kappa$ onto $[N^\kappa . \xi_0^\kappa]$ Then

$$p \in (N^\kappa)' = M^\kappa$$

and the preceding discussion shows that

$$(N^\kappa)_p \text{ is a type } II_1 \text{ factor .}$$

The central support of the projection p is the same as the central support of the orthogonal projection of $\mathcal{H}^\kappa$ onto the subspace

$$[(N^\kappa)' \xi_0^\kappa] = [M^\kappa \xi_0^\kappa] = \mathcal{H}^\kappa ,$$

thus p is a faithful projection in $(N^\kappa)'$. It is known that any induction by a faithful projection is an isomorphism (see[6]) , in particular the canonical map

$$N^\kappa \longrightarrow (N^\kappa)_p$$

is an isomorphism .

Therefore , N^κ is a type II_1 factor . Since $(M^\kappa)' = N^\kappa$, it follows that M^κ is a type II factor . In order to show that M^κ is infinite we shall construct an infinite family of mutually orthogonal and equivalent projections $\{q_n\}$ in M .

Consider the vectors

$$\xi_n = e_{k_1} \otimes \dots \otimes e_{k_{n-1}} \otimes e_{k_n} \otimes e_{k_n} \otimes e_{k_{n+2}} \otimes e_{k_{n+3}} \otimes \dots \in \mathcal{H}^\kappa$$

and denote by q_n the orthogonal projection of $\mathcal{H}^\kappa$ onto $[N^\kappa \xi_n]$. Then

$$q_n \in (N^\kappa)' = M^\kappa \quad .$$

It is clear that for any $n \neq m$ and for any $\sigma, \tau \in S(\infty)$ we have

$$(\nu^\kappa(\sigma)\,\xi_n \mid \nu^\kappa(\tau)\,\xi_m) = 0 \quad ,$$

thus

$$q_n \perp q_m \quad \text{for} \quad n \neq m \quad .$$

Finally, define $U_n \in U(\infty)$ by

$$U_n\, e_j = \begin{cases} e_{k_{n+1}} & \text{if} \quad j = k_n \\ e_{k_{n+2}} & \text{if} \quad j = k_{n+1} \\ e_{k_n} & \text{if} \quad j = k_{n+2} \\ e_j & \text{otherwise} \end{cases}$$

and $\sigma_n \in S(\infty)$ by

$$\sigma_n = \begin{pmatrix} 1 & \dots & n-1 & n & n+1 & n+2 & n+3 & n+4 & \dots \\ 1 & \dots & n-1 & n+2 & n+1 & n & n+3 & n+4 & \dots \end{pmatrix}$$

Then

$$\mu^\kappa(U_n)\,\xi_n = \nu^\kappa(\sigma_n)\,\xi_{n+1} \quad ,$$

so

$$\mu^\kappa(U_n)\, q_n\, \mathcal{H}^\kappa = q_{n+1}\, \mathcal{H}^\kappa \quad .$$

It follows that $\mu^\kappa(U_n)\, q_n$ is a partial isometry in M^κ with initial projection q_n and final projection q_{n+1} .

Thus ,

$$q_n \sim q_{n+1} \quad ,$$

which completes the proof of the Theorem .

Q.E.D.

V.1.6. For any permutation $\sigma \in S(\infty)$ we can define a unitary operator $U_\sigma^\kappa \in U(\infty)$ by

$$U_\sigma^\kappa e_{k_j} = e_{k_{\sigma(j)}} \qquad \text{for each } j \in \mathbb{N} ,$$
$$U_\sigma^\kappa e_k = e_k \qquad \text{if } k \notin \kappa = \{k_j\} .$$

Then

$$\Sigma^\kappa(\infty) = \{U_\sigma^\kappa \in U(\infty) ; \sigma \in S(\infty)\}$$

is a subgroup of $U(\infty)$. Of course, $\Sigma^\kappa(\infty)$ and $S(\infty)$ are isomorphic as abstract groups.

Since

$$\mu^\kappa(U_\sigma^\kappa)\xi_0^\kappa = \nu^\kappa(\sigma^{-1})\xi_0^\kappa ,$$

it follows that

$$\mu^\kappa(\Sigma^\kappa(\infty))[N^\kappa \xi_0^\kappa] = [N^\kappa \xi_0^\kappa] .$$

As we have seen, <u>the von Neumann algebra</u> $(N^\kappa)_p$ <u>is generated by</u>

$$\nu^\kappa(S(\infty)) \Big| [N^\kappa \xi_0^\kappa] .$$

Moreover, its commutant, <u>the von Neumann algebra</u> $(M^\kappa)_p$ <u>is generated by</u>

$$\mu^\kappa(\Sigma^\kappa(\infty)) \Big| [N^\kappa \xi_0^\kappa] .$$

This can be seen as follows. The vectors $\{\nu^\kappa(\sigma)\xi_0^\kappa ; \sigma \in S(\infty)\}$ form an orthonormal basis in $[N^\kappa \xi_0^\kappa]$. Thus we may identify $[N^\kappa \xi_0^\kappa]$ with the Hilbert space $\ell^2(S(\infty))$ in such a way that $\nu^\kappa(\sigma)$ becomes the left translation by σ and $\mu^\kappa(U_\sigma^\kappa)$ becomes the right translation by σ, for every $\sigma \in S(\infty)$. Since the left regular representation of $S(\infty)$ is the commutant of its right regular representation, our assertion is apparent.

V.1.7. The following theorem shows that the representations μ^κ obtained from essentially different sequences κ are non-equivalent .

THEOREM . Let $\kappa = \{k_n\}$, $\kappa' = \{k'_n\}$ be two strictly increasing sequences of positive integers . Then the representations μ^κ and $\mu^{\kappa'}$ of $U(\infty)$ are unitarily equivalent if and only if there exists $n_o \in \mathbb{N}$ such that

$$k_n = k'_n \qquad \text{for all} \quad n \geq n_o \quad .$$

Proof . If this condition is satisfied , then it is obvious that the representations μ^κ and $\mu^{\kappa'}$ are identical .

Suppose now that μ^κ and $\mu^{\kappa'}$ are equivalent . Define the operators $U_n \in U(\infty)$ by

$$U_n\, e_k = \begin{cases} -e_{k_n} & \text{if} \quad k = k_n \quad , \\ e_k & \text{if} \quad k \neq k_n \quad . \end{cases}$$

If $n > m$, we have

$$\mu^\kappa(U_n)\,\xi = -\xi \qquad \text{for each} \quad \xi \in \mathcal{H}^m_{k_m} \subset \mathcal{H}^\kappa \ ,$$

so

$$\lim_{n\to\infty} \mu^\kappa(U_n)\,\xi = -\xi \qquad \text{for each} \quad \xi \in \bigcup_{m=1}^{\infty} \mathcal{H}^m_{k_m} \subset \mathcal{H}^\kappa \ .$$

Since $\overline{\bigcup_{m=1}^{\infty} \mathcal{H}^m_{k_m}} = \mathcal{H}^\kappa$, it follows that

$$\lim_{n\to\infty} \mu^\kappa(U_n)\,\xi = -\xi \qquad \text{for each} \quad \xi \in \mathcal{H}^\kappa \quad .$$

By the equivalence of the representations μ^κ and $\mu^{\kappa'}$ we infer

$$\lim_{n\to\infty} \mu^{\kappa'}(U_n)\,\xi_o^{\kappa'} = -\xi_o^{\kappa'} \qquad \text{where} \quad \xi_o^{\kappa'} = \bigotimes_{j=1}^{\infty} e_{k'_j} \in \mathcal{H}^{\kappa'} \ .$$

But it is clear that

$$k_n \notin \{k_1', k_2', \dots, k_j', \dots\} \Longrightarrow \mu^{\kappa'}(U_n)\,\xi_0^{\kappa'} = \xi_0^{\kappa'} .$$

Therefore, there exists $m_0 \in \mathbb{N}$ such that

$$\{k_n\}_{n \geq m_0} \subset \{k_n'\}_{n \geq 1} .$$

By a dual argument we find $m_0' \in \mathbb{N}$ such that

$$\{k_n'\}_{n \geq m_0'} \subset \{k_n\}_{n \geq 1} .$$

Changing, if necessary, the number $m_0 \in \mathbb{N}$ we may suppose that

$$\{k_n\}_{n \geq m_0} \subset \{k_n'\}_{n \geq m_0'} .$$

On the other hand, there exists $n_0' \geq m_0'$ such that

$$\{k_n'\}_{n \geq n_0'} \subset \{k_n\}_{n \geq m_0} .$$

Choose $n_0 \geq m_0$ such that $k'_{n_0'} = k_{n_0}$. Then it is easy to see that

$$\{k_n'\}_{n \geq n_0'} = \{k_n\}_{n \geq n_0} .$$

It remains to show that

$$n_0' = n_0 .$$

Suppose the contrary holds, for instance

$$n_0' = n_0 + r \quad , \quad r \in \mathbb{N} ,$$

choose $\lambda \in \mathbb{C}$, $|\lambda| = 1$, such that

$$\lambda^r \neq 1$$

and define the operators $V_n \in U(\infty)$ by

$$V_n e_k = \begin{cases} \lambda e_k & \text{if} \quad 1 \leq k \leq k_n , \\ e_k & \text{if} \quad k > k_n . \end{cases}$$

If $n > m$ we have

$$(\mu^\kappa(V_n)\xi \mid \xi) = \lambda^n \|\xi\|^2 \qquad \text{for each} \quad \xi \in \mathcal{H}^m_{k_m} \subset \mathcal{H}^\kappa ,$$

so

$$\lim_{n \to \infty} \frac{(\mu^\kappa(V_n)\xi \mid \xi)}{\lambda^n} = \|\xi\|^2 \qquad \text{for each} \quad \xi \in \mathcal{H}^\kappa .$$

By the equivalence of the representations μ^{κ} and $\mu^{\kappa'}$ we infer

(1) $$\lim_{n\to\infty} \frac{(\mu^{\kappa'}(V_n)\xi'\mid\xi')}{\lambda^n} = \|\xi'\|^2 \quad \text{for each} \quad \xi' \in \mathcal{H}^{\kappa'} .$$

On the other hand, if $n > m$ we have

$$(\mu^{\kappa'}(V_n)\xi'\mid\xi') = \lambda^{n+r}\|\xi'\|^2 \quad \text{for each} \quad \xi' \in \mathcal{H}^m_{k_m} \subset \mathcal{H}^{\kappa'},$$

so

(2) $$\lim_{n\to\infty} \frac{(\mu^{\kappa'}(V_n)\xi'\mid\xi')}{\lambda^n} = \lambda^r\|\xi'\|^2 \quad \text{for each} \quad \xi' \in \mathcal{H}^{\kappa'} .$$

[Note that $k_m > k_{m-r} = k'_m$ for m large, so $\mathcal{H}^m_{k_m} \supset \mathcal{H}^m_{k'_m}$ as subspaces of $\mathcal{H}^{\kappa'}$ and this implies

$$\mathcal{H}^{\kappa'} = \overline{\bigcup_{m=1}^{\infty} \mathcal{H}^m_{k_m}} .]$$

Comparing (1) and (2) we get

$$\lambda^r = 1$$

and this is a contradiction.

Q.E.D.

V.1.8. For every $U \in U(\infty)$ we have

$$(\mu^{\kappa}(U)\,\xi^{\kappa}_0 \mid \xi^{\kappa}_0) = \prod_{j=1}^{\infty} (Ue_{k_j} \mid e_{k_j})$$

Therefore, <u>the function</u>

(3) $$\varphi^{\kappa}(U) = \prod_{j=1}^{\infty} (Ue_{k_j} \mid e_{k_j}) \quad ; \quad U \in U(\infty)$$

<u>is of positive type on</u> $U(\infty)$ and μ^{κ} is the representation of $U(\infty)$ associated to φ^{κ}.

We shall show that φ^{κ} <u>is uniformly continuous with respect to the metric of</u> $U_1(\infty)$,

$$d(U', U'') = \mathrm{Tr}\,|U' - U''| \quad , \quad U', U'' \in U_1(\infty)$$

Since this metric is both left and right invariant and since φ^{κ} is of positive type (recall that this implies

$$\left|\varphi^{\kappa}(U') - \varphi^{\kappa}(U'')\right|^2 \leqslant 2\left|1 - \varphi^{\kappa}((U')^{-1}U'')\right| ,$$

see [7], 13.4.7.) it is sufficient to prove the continuity of φ^{κ} in the identity $I \in U(\infty)$. First remark that for any normal operator T on H and for any vector $\xi \in H$ we have

$$\begin{aligned} \left|(T\xi \mid \xi)\right| &\leq \left|((\mathrm{Re}\, T)\xi \mid \xi)\right| + \left|((\mathrm{Im}\, T)\xi \mid \xi)\right| \\ &\leq (\left|\mathrm{Re}\, T\right| \xi \mid \xi) + (\left|\mathrm{Im}\, T\right| \xi \mid \xi) \\ &\leq 2\,(\left|T\right| \xi \mid \xi) . \end{aligned}$$

Thus , we obtain

$$\begin{aligned} \left|\varphi^{\kappa}(U) - 1\right| &= \left|\prod_{j=1} (Ue_{k_j} \mid e_{k_j}) - 1\right| \\ &\leqslant \prod_{j=1}^{\infty}\left(1 + \left|(Ue_{k_j} \mid e_{k_j}) - 1\right|\right) - 1 \\ &\leqslant \exp\left(\sum_{j=1}^{\infty} \left|(Ue_{k_j} \mid e_{k_j}) - 1\right|\right) - 1 \\ &= \exp\left(\sum_{j=1}^{\infty} \left|((U - I)e_{k_j} \mid e_{k_j})\right|\right) - 1 \\ &\leqslant \exp\left(2\sum_{j=1}^{\infty} (\left|U - I\right| e_{k_j} \mid e_{k_j})\right) - 1 \\ &\leqslant \exp\left(2\,\mathrm{Tr}\left|U - I\right|\right) - 1 \end{aligned}$$

It follows that φ^{κ} extends to a uniformly continuous function of positive type on $U_1(\infty)$, denoted also by φ^{κ}.Consequently, <u>the representation μ^{κ} of</u> $U(\infty)$ <u>extends to a representation of</u> $U_1(\infty)$, denoted again by μ^{κ} .

It is easy to see that formula (3) extends to $U_1(\infty)$ and that for decomposable vectors $\bigotimes_{j=1}^{\infty} \xi_j \in \mathcal{H}^{\kappa}$ and $U \in U_1(\infty)$ we have

$$\bigotimes_{j=1}^{\infty} U\xi_j \in \mathcal{H}^{\kappa} \quad \text{and} \quad \mu^{\kappa}(U)\left(\bigotimes_{j=1}^{\infty} \xi_j\right) = \bigotimes_{j=1}^{\infty} U\xi_j \quad .$$

Since $U(\infty)$ is dense in $U_1(\infty)$, the von Neumann algebra generated in $L(\mathcal{H}^{\kappa})$ by $\mu^{\kappa}(U_1(\infty))$ is equal to the von Neumann algebra M^{κ} generated by $\mu^{\kappa}(U(\infty))$. Hence μ^{κ} _is a type_ II_{∞} _factor representation of_ $U_1(\infty)$. Moreover , _the equivalence of_ μ^{κ} _and_ $\mu^{\kappa'}$ _as representations of_ $U_1(\infty)$ _reduces to their equivalence as representations of_ $U(\infty)$.

V.1.9. Let V be a unitary operator on H . Then

$$i_V : U_1(\infty) \ni U \longrightarrow V^*UV \in U_1(\infty)$$

is an automorphism of $U_1(\infty)$. Therefore

$$\mu^{\kappa} \circ i_V$$

is a type II_{∞} factor representation of $U_1(\infty)$ and

$$\mu^{\kappa} \circ i_V \mid U(\infty)$$

is a type II_{∞} factor representation of $U(\infty)$.

It is clear that $\mu^{\kappa} \circ i_V \mid U(\infty)$ is the natural representation of $U(\infty)$ on the Hilbert space infinite tensor product of a sequence of copies of H along the sequence of vectors

$$(Ve_{k_1} , Ve_{k_2} , \dots , Ve_{k_n} , \dots)$$

Thus , we obtain the following result .

Consider an arbitrary orthonormal system α in H ,

$$\alpha = (a_1 , a_2 , \dots , a_n , \dots)$$

and define the Hilbert space

$$\mathcal{H}^{\alpha}$$

as the infinite tensor product of a sequence of copies of H along the sequence α. There is a natural representation

$$\mu^{\alpha}$$

of $U(\infty)$ on $\mathcal{H}^{\alpha}$ such that

$$\mu^{\alpha}(U)\left(\bigotimes_{j=1}^{\infty}\xi_j\right) = \bigotimes_{j=1}^{\infty} U\xi_j \quad ; \quad U \in U(\infty)$$

for all decomposable vectors

$$\bigotimes_{j=1}^{\infty}\xi_j \in \mathcal{H}^{\alpha} \quad .$$

THEOREM . <u>For any orthonormal system</u> $\alpha = \{a_n\} \subset H$, <u>the representation</u> μ^{α} <u>of</u> $U(\infty)$ <u>on</u> $\mathcal{H}^{\alpha}$ <u>is a factor representation of type</u> II_{∞} .

This extends Theorem V.1.5. There is also an extension of Theorem V.1.4. , namely <u>the commutant of</u> μ^{α} <u>is generated by the natural representation of</u> $S(\infty)$ <u>on</u> $\mathcal{H}^{\alpha}$.

Note that μ^{α} is the representation of $U(\infty)$ associated to the function of positive type

$$\varphi^{\alpha}(U) = \prod_{j=1}^{\infty}(Ua_j \mid a_j) \quad ; \quad U \in U(\infty) .$$

V.1.10. Now the following <u>problem</u> arises :

<u>Given two orthonormal systems</u> $\alpha = \{a_n\}$, $\beta = \{b_n\}$ <u>in</u> H , <u>find necessary and sufficient conditions in order that the representations</u> μ^{α} <u>and</u> μ^{β} <u>be equivalent</u> .

Theorem V.1.7. contains the answer to this problem in a particular case .

In general , if there is an <u>arbitrary</u> permutation of $\mathbb{N}$ such that

$$b_n = \theta_n a_{\sigma(n)} \quad \text{for suitable} \quad \theta_n \in \mathbb{C} \ , |\theta_n| = 1 \ ,$$

then the functions of positive type φ^α and φ^β are equal , so the representations μ^α and μ^β are equivalent .

On the other hand , suppose there exists an operator $U \in U_1(\infty)$ such that

$$b_n = Ua_n \quad .$$

Then the Hilbert spaces $\mathcal{H}^\alpha$ and $\mathcal{H}^\beta$ coincide , so the representations μ^α and μ^β are identical .

A reasonable <u>conjecture</u> might be that

<u>The representations</u> μ^α <u>and</u> μ^β <u>are unitarily equivalent if and only if there exist</u>

<u>a permutation</u> σ <u>of</u> $\mathbb{N}$

<u>an operator</u> $U \in U_1(\infty)$

<u>such that</u>

$$U b_n = \theta_n a_{\sigma(n)} \quad \underline{\text{for suitable}} \quad \theta_n \in \mathbb{C} \ , |\theta_n| = 1 \ .$$

A weak result in this direction is the following

PROPOSITION . <u>Let</u> $\alpha = \{a_n\}$, $\beta = \{b_n\}$ <u>be orthonormal systems in</u> H . <u>Suppose that the representations</u> μ^α <u>and</u> μ^β <u>of</u> $U(\infty)$ <u>are equivalent</u> . <u>Then there are finite sets</u> $F_\alpha \subset \mathbb{N}$, $F_\beta \subset \mathbb{N}$ <u>and a bijective map</u>

$$\sigma : \mathbb{N} \setminus F_\beta \longrightarrow \mathbb{N} \setminus F_\alpha$$

<u>such</u> <u>that</u>

$$\lim_{n \to \infty} \|b_n - \theta_n \, a_{\sigma(n)}\| = 0 \quad \underline{\text{for suitable}} \quad \theta_n \quad \mathbb{C} \quad , |\theta_n| = 1 \; .$$

<u>Proof</u> . Let $\{c_n\}$ be an orthonormal basis of H which contains the orthonormal system $\{a_n\}$ such that

$$a_n = c_{k_n}$$

where $\{k_n\}$ is a strictly increasing sequence of positive integers.

Define the operators $U_n \in U_1(\infty)$ by

$$U_n \, c_k = \begin{cases} -c_{k_n} & \text{if} \quad k = k_n \quad , \\ c_k & \text{if} \quad k \neq k_n \quad . \end{cases}$$

As in the proof of Theorem V.1.7. we see that for each $\xi \in \mathcal{H}^\alpha$

$$\lim_{n \to \infty} (\mu^\alpha(U_n) \xi \mid \xi) = - \|\xi\|^2 \qquad .$$

Since μ^β is equivalent to μ^α, we obtain

$$\lim_{n \to \infty} \left(\mu^\beta(U_n)\left(\bigotimes_{j=1}^{\infty} b_j\right) \Big| \left(\bigotimes_{j=1}^{\infty} b_j\right)\right) = -1 \qquad ,$$

that is

(4) $$\lim_{n \to \infty} \prod_{j=1}^{\infty} (U_n \, b_j \mid b_j) = -1 \qquad .$$

Write

$$b_j = \sum_{k=1}^{\infty} \gamma_j^k \, c_k \qquad \text{where} \qquad \gamma_j^k = (b_j \mid c_k) \; .$$

Then the relation (4) becomes

(5) $$\lim_{n \to \infty} \prod_{j=1}^{\infty} (1 - 2|\gamma_j^{k_n}|^2) = -1 \qquad .$$

Since

$$\sum_{j=1}^{\infty} |\gamma_j^{k_n}|^2 \leq 1 \qquad ,$$

it follows that for each $n \in \mathbb{N}$ there is at most one index $j \in \mathbb{N}$ such that

$$1 - 2\left|\gamma_j^{k_n}\right|^2 < 0 .$$

Owing to relation (5) we infer the existence of a positive integer n_0 with the following property :

for every $n \geq n_0$ there is a unique $j(n) \in \mathbb{N}$ such that

$$1 - 2\left|\gamma_{j(n)}^{k_n}\right|^2 < 0 .$$

Since

$$\left|1 - 2\left|\gamma_j^{k_n}\right|^2\right| = \left|(U_n b_j \mid b_j)\right| \leq 1 ,$$

using again the relation (5) we obtain

$$\lim_{n \to \infty} (1 - 2\left|\gamma_{j(n)}^{k_n}\right|^2) = -1 . \tag{6}$$

Since

$$\sum_{k=1}^{\infty} \left|\gamma_j^k\right|^2 = 1 ,$$

the map

$$\left\{n \in \mathbb{N} ; n \geq n_0\right\} \ni n \longrightarrow j(n) \in \mathbb{N}$$

is injective . The relation (6) rewrites

$$\lim_{n \to \infty} \left|(b_{j(n)} \mid a_n)\right| = 1 .$$

By a dual argument we find $m_0 \in \mathbb{N}$ and an injective map

$$\left\{n \in \mathbb{N} ; n \geq m_0\right\} \ni n \longrightarrow i(n) \in \mathbb{N}$$

such that

$$\lim_{n \to \infty} \left|(b_n \mid a_{i(n)})\right| = 1 .$$

From these it follows easily that for sufficiently large $n \in \mathbb{N}$ we have

$$i(j(n)) = n \quad .$$

Therefore there are finite sets $F_\alpha \subset \mathbb{N}$, $F_\beta \subset \mathbb{N}$ and a bijective map (namely a restriction of $n \longrightarrow i(n)$)

$$\sigma : \mathbb{N} \setminus F_\beta \longrightarrow \mathbb{N} \setminus F_\alpha$$

such that

$$\lim_{n\to\infty} \left|(b_n \mid a_{\sigma(n)})\right| = 1 \quad .$$

Define , for $n \in \mathbb{N} \setminus F_\beta$,

$$\theta_n = \frac{(b_n \mid a_{\sigma(n)})}{|(b_n \mid a_{\sigma(n)})|} \quad .$$

Then

$$0 < (b_n \mid \theta_n a_{\sigma(n)}) \xrightarrow[n\to\infty]{} 1$$

and clearly

$$\lim_{n\to\infty} \| b_n - \theta_n a_{\sigma(n)} \| = 0 \quad .$$

Q.E.D.

Suppose that α and β are orthonormal bases in H . In order to prove the above conjecture in this case , it would be sufficient to show that

$$\text{card } F_\alpha = \text{card } F_\beta$$

and

$$\sum_{n=1}^{\infty} \left(1 - (b_n \mid \theta_n a_{\sigma(n)})\right) < +\infty \quad ,$$

where the notation is as in the preceding Proposition .

§ 2 Other type II_∞ factor representations

In this section we shall briefly sketch the construction of another kind of type II_∞ factor representations of $U(\infty)$.

The notations and the results from Chapters I , II , III will be freely used . We shall often identify a representation $\rho_n \in \widehat{U(n)}$ with its signature . For example , $p_{(m_1,\ldots,m_n)}$ stands for the projection $p_{\rho_n} \in A = A(U(\infty))$ where $\rho_n \in \widehat{U(n)}$ corresponds to the signature $(m_1,\ldots,m_n)$.

V.2.1. Let $\omega \subset \Omega$ be the closure of Γ- orbit corresponding to the upper signature $(\infty, 1 , 0 , 0 , \ldots)$ and to the lower signature $(0 , 0 , 0 , \ldots)$. The points of ω are the symbols

$$t = (\rho_1(t) \prec \rho_2(t) \prec \ldots \prec \rho_j(t) \prec \ldots)$$

such that $\rho_j(t) \in \widehat{U(j)}$ is either of the form

$$(m(t),1,\underbrace{0,\ldots\ldots,0}_{(j-2)\text{-times}})$$

or of the form

$$(m(t),\underbrace{0,0,\ldots\ldots,0}_{(j-1)\text{-times}})$$

Let further $\omega' \subset \omega$ consist of those points $t \in \omega$ with $n_0(t) = +\infty$ such that for large enough j (depending on t) the signature of $\rho_j(t)$ is of the form $(m,1,0,\ldots,0)$. Clearly , ω' is a Γ - invariant Borel subset of ω . Let us also denote

$$\Theta_{(\rho_1 \prec \cdots \prec \rho_n)} = \left\{ t \in \omega' ;\ \rho_j(t) = \rho_j \quad \text{for} \quad j = 1,\ldots,n \right\} .$$

Consider now a system of positive real numbers

$$\left\{ c_m^{(n)} \right\}_{m,n \in \mathbb{N}}$$

such that

$$(1) \qquad c_m^{(n)} = \sum_{p \geqslant m} c_p^{(n+1)} \quad ; \quad m,n \in \mathbb{N} \quad .$$

For a given system of numbers $\{c_m^{(n)}\}$ satisfying (1), there is a unique Γ- invariant, sigma-finite Borel measure μ on ω such that:

$$\text{(i)} \quad \mu(\theta_{(\rho_1 < \dots < \rho_n)}) = c_m^{(n)} \quad \text{if} \quad \rho_n = (m,1,0,\dots,0) ,$$

$$\text{(ii)} \quad \mu(\omega \setminus \omega') = 0 .$$

Moreover, any Γ- invariant Borel measure μ' on ω, such that

$$0 \leqslant \mu' \leqslant \mu$$

is of the same kind, i.e. defined as above by a system of numbers $\{c'^{(n)}_m\}$ satisfying the conditions (1) and such that

$$0 \leqslant c'^{(n)}_m \leqslant c_m^{(n)} \qquad ; \quad m,n \in \mathbb{N} .$$

On the other hand, consider the closure of Γ- orbit ω_1 corresponding to the upper signature $(\infty, 0, 0, \dots)$ and to the lower signature $(0, 0, 0, \dots)$ and the subset ω_1' of ω_1 consisting of those points $t \in \omega_1$ with $n_0(t) = +\infty$. Consider also

$$\theta^1_{(\rho_1 < \dots < \rho_n)} = \{t \in \omega_1' ;\ \rho_j(t) = \rho_j \ \text{ for } \ j = 1,\dots,n\} .$$

Then for every system $\{c_m^{(n)}\}$ satisfying (1) there is a unique Γ- invariant, finite Borel measure ν on ω_1 such that

$$(\mathrm{i}_1) \quad \nu(\theta^1_{(\rho_1 < \dots < \rho_n)}) = c_{m+1}^{(n)} \quad \text{if} \quad \rho_n = (m,0,0,\dots,0) ,$$

$$(\mathrm{ii}_1) \quad \nu(\omega_1 \setminus \omega_1') = 0 .$$

Also, if ν' is a Γ- invariant Borel measure on ω_1 such that

$$0 \leqslant \nu' \leqslant \nu$$

then ν' also corresponds to a system of numbers $\{c'^{(n)}_m\}$ satisfying the conditions (1) and such that

$$0 \leqslant c'^{(n)}_m \leqslant c^{(n)}_m \qquad ; m,n \in \mathbb{N} \qquad .$$

These remarks show that, for a system of positive numbers $\{c^{(n)}_m\}$ satisfying (1), the measure μ on ω it defines is Γ- ergodic if and only if the other measure ν on ω_1 it defines is Γ- ergodic.

V.2.2. Now a finite Γ- invariant, Γ- ergodic Borel measure on ω_1 yields a type II_1 factor representation of $U(\infty)$ and conversely, such measures arise from type II_1 factor representations of $U(\infty)$ the restrictions of which to the $U(n)$'s decompose only according to representations of signatures $(m,0,0,\ldots,0)$.

The character of the type II_1 factor representation corresponding to ν is then given by :

$$\chi \mid U(n) = \sum_{m=0}^{\infty} c^{(n)}_{m+1} \; \chi_{(m,\underbrace{0,\ldots\ldots,0}_{(n-1)\text{-times}})}$$

It is known that the functions

$$V \longmapsto \det((1-b)(I-bV)^{-1}) \quad , \quad 0<b<1 \quad ,$$

on $U(\infty)$ are characters of type II_1 factor representations of $U(\infty)$ (see[ ]) and for $V \in U(n)$ we have

$$\det((1-b)(I-bV)^{-1}) = \sum_{m=0}^{\infty} b^m (1-b)^n \chi_{(m,\underbrace{0,\ldots\ldots,0}_{(n-1)\text{-times}})}(V)$$

The last equality can be easily established by computations on a maximal torus of $U(n)$.

Thus, <u>the measure μ on ω corresponding to the system</u>

$$c_m^{(n)} = b^{m-1}(1 - b)^n \quad , \quad 0 < b < 1 \quad ,$$

<u>is</u> Γ - <u>ergodic</u> .

V.2.3. The measure μ on ω defined by these numbers $c_m^{(n)}$ <u>is sigma-finite</u> indeed , <u>but it is not finite</u> .

This can be proved as follows . For $p \geqslant 1$ we have

$$\mu(\Theta_{(\rho_1 < \cdots < \rho_{n-1} < (p,\underbrace{0,\ldots\ldots,0}_{(n-1)\text{-times}}))}) =$$

$$= \sum_{q \geqslant p} c_q^{(n+1)} + \sum_{q \geqslant p} \mu(\Theta_{(\rho_1 < \cdots < \rho_{n-1} < (p,\underbrace{0,\ldots\ldots,0}_{(n-1)\text{-times}}) < (q,\underbrace{0,\ldots,0}_{n\text{-times}}))})$$

$$\geqslant \sum_{q \geqslant p} b^{q-1}(1 - b)^{n+1} = b^{p-1}(1 - b)^n \quad .$$

Suppose now we have proved

$$\mu(\Theta_{(\rho_1 < \cdots < \rho_{n-1} < (p,\underbrace{0,\ldots\ldots,0}_{(n-1)\text{-times}}))}) \geqslant k\, b^{p-1}(1 - b)^n \quad ,$$

for all $p \geqslant 1$ and $n \geqslant 1$. Then , using the same equality as above , we have

$$\mu(\Theta_{(\rho_1 < \cdots < \rho_{n-1} < (p,\underbrace{0,\ldots\ldots,0}_{(n-1)\text{-times}}))}) \geqslant$$

$$\geqslant b^{p-1}(1 - b)^n + \sum_{q \geqslant p} k\, b^{q-1}(1 - b)^{n+1}$$

$$= (k + 1)\, b^{p-1}(1 - b)^n \quad .$$

Thus , we must have

$$\mu(\Theta_{(\rho_1 < \cdots < \rho_{n-1} < (p,\underbrace{0,\ldots\ldots,0}_{(n-1)\text{-times}}))}) = + \infty \quad .$$

V.2.4. By the general theory, the Γ- ergodic, Γ- invariant, sigma-finite measure μ on $\omega \subset \Omega$ defines a type II_∞ factor representation π_μ of $U(\infty)$. What singles this factor representation out up to quasi-equivalence ? The answer is the following:

Given a semifinite representation ρ of $U(\infty)$, it is quasi-equivalent to π_μ if and only if

(2) $$\mathrm{Tr}\,(\rho(p_{(m,1,\underbrace{0,\ldots\ldots,0}_{(n-2)\text{-times}})})) = d_{(m,1,\underbrace{0,\ldots\ldots,0}_{(n-2)\text{-times}})}\, b^{m-1}(1-b)^n$$

(3) $$\bigvee_{m,n} \rho(p_{(m,1,\underbrace{0,\ldots\ldots,0}_{(n-2)\text{-times}})}) = I \quad .$$

What we have to show is that two representations satisfying these conditions are quasi-equivalent. This will follow from usual arguments about (unimodular) Hilbert algebras if we make the following remarks.

The condition (3) implies that

$$\sum_{m,n} \rho(B_{(m,1,\underbrace{0,\ldots\ldots,0}_{(n-2)\text{-times}})})$$

is weakly dense in $(\rho(U(\infty)))''$ and, since $\rho(B_{(m,1,0,\ldots,0)})$ are finite dimensional factors, the trace is completely determined on the above finite-dimensional subspace by the condition (2).

Moreover

$$\sum_{m,n} \rho(B_{(m,1,\underbrace{0,\ldots\ldots,0}_{(n-2)\text{-times}})})$$

is a $*$ - subalgebra of $(\rho(U(\infty)))''$ and it is $*$ - isomorphic with the $*$ - subalgebra

$$\sum_{m,n} B_{(m,1,\underbrace{0,\ldots\ldots,0}_{(n-2)\text{-times}})}$$

of $A = A(U(\infty))$.

V.2.5. With these preparations we can now exhibit another realization of the representations π_μ .

Let ρ^1 be the natural representation of $U(\infty)$ on its underlying Hilbert space H with orthonormal basis $\{e_n\}_{n \in \mathbb{N}}$.

Consider also ρ^2 a type II_1 factor representation of $U(\infty)$ of character

$$\det\left((1 - b)(I - bU)^{-1}\right)$$

in standard form on a Hilbert space K and $\xi \in K$ a trace vector.

Then $(\rho^1(U(\infty)) \otimes \rho^2(U(\infty)))''$ is a type II_∞ factor acting on $H \otimes K$, the trace being defined by the weight

$$\varphi = b^{-1} \sum_{n=1}^{\infty} \omega_{e_n \otimes \xi} .$$

Let further ρ^3 be the tensor product of the representations ρ^1 and ρ^2 :

$$\rho^3(U) = \rho^1(U) \otimes \rho^2(U) , \quad U \in U(\infty) .$$

The projections

$$\rho^3(p_{(m,1,\underbrace{0,\ldots\ldots,0}_{(n-2)\text{-times}})})$$

are finite projections in $(\rho^1(U(\infty)) \otimes \rho^2(U(\infty)))''$. Indeed ,

$$\rho^3(p_{(m,1,\underbrace{0,\ldots\ldots,0}_{(n-2)\text{-times}})}) \in (\rho^1(U(n)) \otimes \rho^2(U(n)))''$$

and thus

$$\varphi(\rho^3(p_{(m,1,\underbrace{0,\ldots\ldots,0}_{(n-2)\text{-times}})})) =$$

$$= b^{-1}\sum_{j=1}^{n}\omega_{e_j\otimes\xi}\;(\rho^3(p_{(m,1,\underbrace{0,\ldots\ldots,0}_{(n-2)\text{-times}})})) \quad ,$$

since

$$H \otimes K = (H_n \otimes K) \oplus ((H \ominus H_n) \otimes K)$$

and the representation of $U(n)$ on $(H \ominus H_n) \otimes K$, being quasi - equivalent to $\rho^2 \mid U(n)$, contains only signatures of the form $(m,0,0,\ldots,0)$.

Now , we have

$$b^{-1}\sum_{j=1}^{n}\omega_{e_j\otimes\xi}\circ\rho^3 \,\big|\, U(n) =$$

$$= b^{-1}\chi_{(1,\underbrace{0,\ldots\ldots,0}_{(n-1)\text{-times}})}\;\chi_{\rho^2}\,\big|\,U(n) =$$

$$= b^{-1}\chi_{(1,\underbrace{0,\ldots\ldots,0}_{(n-1)\text{-times}})}\;\sum_{m=0}^{\infty} b^m(1-b)^n\,\chi_{(m,\underbrace{0,\ldots\ldots,0}_{(n-1)\text{-times}})} =$$

$$= b^{-1}(1-b)^n\chi_{(1,\underbrace{0,\ldots\ldots,0}_{(n-1)\text{-times}})} +$$

$$+ b^{-1}\sum_{m=1}^{\infty} b^m(1-b)^n\Big[\chi_{(m+1,\underbrace{0,\ldots\ldots,0}_{(n-1)\text{-times}})} + \chi_{(m,1,\underbrace{0,\ldots\ldots,0}_{(n-2)\text{-times}})}\Big]$$

where we have used the known fact that

$$\chi_{(1,0,0,\ldots,0)}\,\chi_{(m,0,0,\ldots,0)} = \chi_{(m+1,0,0,\ldots,0)} + \chi_{(m,1,0,\ldots,0)}$$

for $m \geq 1$.

Hence we have

$$\varphi(\rho^3(p_{(m,1,\underbrace{0,\ldots\ldots,0}_{(n-2)\text{-times}})})) = b^{m-1}(1-b)^n d_{(m,1,\underbrace{0,\ldots\ldots,0}_{(n-2)\text{-times}})} .$$

Consider

$$p^{(n)} = \sum_{m=1}^{\infty} \rho^3(p_{(m,1,\underbrace{0,\ldots\ldots,0}_{(n-2)\text{-times}})})$$

Then we have

$$p^{(2)} \leqslant p^{(3)} \leqslant p^{(4)} \leqslant \ldots .$$

and

$p^{(n)}(H \otimes K)$ is an invariant subspace for $\rho^3(U(n))$.

It follows that

$$L = \bigvee_{n=2}^{\infty} p^{(n)}(H \otimes K)$$

is an invariant subspace for $\rho^3(U(\infty))$.

It is now easy to see that the restriction of ρ^3 to L satisfies the conditions (2) and (3) .

Thus , <u>the restriction of ρ^3 to L is a type II_∞ factor representation of $U(\infty)$, quasi-equivalent to π_μ</u>.

V.2.6. The problem which naturally arises in connection with these type II_∞ factor representations of $U(\infty)$ is to study the tensor products of irreducible representations of $U(\infty)$ considered by I.Segal ([30]) and A.Kirillov ([21]) and of type II_1 factor representations ([35]).

APPENDIX : IRREDUCIBLE REPRESENTATIONS OF U(n)

As we already mentioned (III.1.1.), there is a bijection between equivalence classes of irreducible representations of U(n) and decreasing n-tuples

$$m_1 \geqslant m_2 \geqslant \dots \geqslant m_n$$

of integers (the "signatures" of irreducible representations).

Below we shall describe for each signature

$$m_1 \geqslant m_2 \geqslant \dots \geqslant m_n$$

an irreducible representation of U(n) in the corresponding equivalence class. In order to avoid mixed tensors, we first consider only positive signatures (i.e. $m_n \geqslant 0$) and then we indicate a way to obtain also representations for the remaining signatures, starting from the positive ones.

For a positive signature

$$m_1 \geqslant m_2 \geqslant \dots \geqslant m_n \geqslant 0$$

consider the following diagram (Young's diagram) :

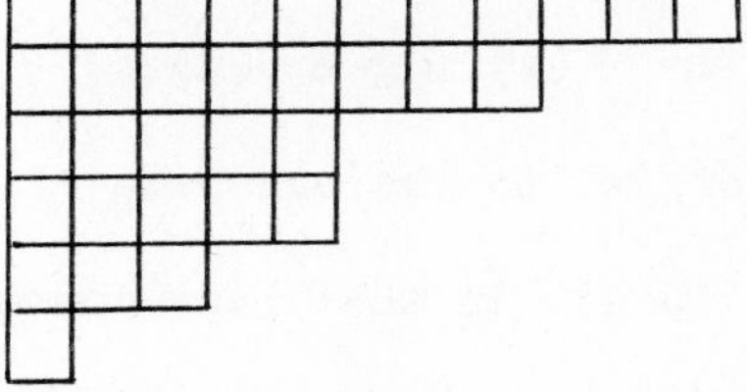

the rows of which have lenght m_1 , m_2 , ... , m_n respectively and insert in squares the numbers $\{1,2,3,\dots,m\}$,$m = m_1 + m_2 + \dots + m_n$, filling first the first column, then the second column and so on. For instance, if n = 3 the signature (4,3,1) yields :

1	4	6	8
2	5	7	
3			

Consider $S(m)$ the group of permutations of the set $\{1, 2, \dots, m\}$ and P, Q the subgroups consisting of those permutations which conserve the rows of the Young diagram, respectively its columns (horizontal and vertical permutations). Denote by $\varepsilon(\sigma)$ the sign of the permutation $\sigma \in S(m)$.

Let H_n be the n-dimensional Hilbert space on which $U(n)$ acts and ρ be the natural representation of $U(n)$ on

$$\mathcal{H}_n^m = \underbrace{H_n \otimes \dots \otimes H_n}_{m\text{-times}} .$$

Consider also the representation π of $S(m)$ on $\mathcal{H}_n^m$ such that :

$$\pi(\sigma)\left(\bigotimes_{j=1}^{m} \xi_j\right) = \bigotimes_{j=1}^{m} \xi_{\sigma^{-1}(j)} \quad ; \xi_1, \dots, \xi_m \in H_n \quad , \quad \sigma \in S(m)$$

and define the linear map $R : \mathcal{H}_n^m \longrightarrow \mathcal{H}_n^m$ by

$$R = \sum_{(p,q) \in P \times Q} \varepsilon(q)\, \pi(qp) .$$

Then R commutes with the $\rho(g)$, $g \in U(n)$ and $R(\mathcal{H}_n^m)$ is an invariant subspace for the $\rho(g)$, $g \in U(n)$.

The restriction of ρ to the subspace $R(\mathcal{H}_n^m)$ is an irreducible representation of $U(n)$ in the class corresponding to the signature $(m_1, m_2, \dots, m_n)$.

For a general signature

$$m_1 \geqslant m_2 \geqslant \dots \geqslant m_n$$

let $\rho(m_1', \dots, m_n')$ be a representation in the class of the signature

$$m_1' \geq m_2' \geq \dots \geq m_n' \geq 0$$

where $m_j' = m_j - m_n$ and then consider the representation

$$U(n) \ni g \longmapsto (\det(g))^{m_n} \rho_{(m_1',\dots,m_n')}(g) \quad .$$

This is a representation in the class corresponding to $(m_1, m_2, \dots, m_n)$.

Let us consider a few particular cases :

a) Suppose $(m_1,\dots,m_n) = (m,0,\dots,0)$.

Then the subspace $R(\mathcal{H}_n^m)$ of $\mathcal{H}_n^m$ is just the space of symmetric tensors, i.e. the space of those $\eta \in \mathcal{H}_n^m$ such that

$$\Pi(\sigma)\eta = \eta \qquad \text{for every} \quad \sigma \in S(m) \; .$$

b) Suppose $(m_1,\dots,m_n) = (\underbrace{1,\dots,1}_{k\text{-times}},\underbrace{0,\dots,0}_{(n-k)\text{-times}})$.

Then $R(\mathcal{H}_n^k)$ is the subspace of those $\eta \in \mathcal{H}_n^k$ such that

$$\Pi(\sigma)\eta = \varepsilon(\sigma)\eta \text{ for every } \sigma \in S(k) \; .$$

That is, $R(\mathcal{H}_n^k)$ is the space of antisymmetric tensors of degree k.

c) Suppose $(m_1,\dots,m_n) = (d,\dots,d)$.

Then the corresponding representation is one-dimensional :

$$U(n) \ni g \longmapsto (\det(g))^d \; .$$

A fundamental result concerning irreducible representations of $U(n)$ is <u>the character formula</u> .

Let $m_1 \geq m_2 \geq \dots \geq m_n$ be a signature. Then

$$\frac{\begin{vmatrix} z_1^{m_1+(n-1)} & z_2^{m_1+(n-1)} & \cdot & z_n^{m_1+(n-1)} \\ z_1^{m_2+(n-2)} & z_2^{m_2+(n-2)} & \cdot & z_n^{m_2+(n-2)} \\ \cdot & \cdot & \cdot & \cdot \\ z_1^{m_n} & z_2^{m_n} & \cdot & z_n^{m_n} \end{vmatrix}}{\prod_{i<j} (z_i - z_j)}$$

is a polinomial in $z_1 , \ldots , z_n , z_1^{-1}, \ldots , z_n^{-1}$, we shall denote by

$$\chi_{(m_1,\ldots,m_n)}(z_1,\ldots,z_n) .$$

Consider $\rho_{(m_1,\ldots,m_n)}$ an irreducible representation of $U(n)$ corresponding to the signature $(m_1,\ldots,m_n)$. Then the character formula can be written as follows :

$$\operatorname{Tr} \rho_{(m_1,\ldots,m_n)}(g) = \chi_{(m_1,\ldots,m_n)}(\gamma_1,\ldots,\gamma_n)$$

where $\gamma_1 , \ldots , \gamma_n$ are the eigenvalues of $g \in U(n)$.

Let us mention that the decomposition of the restriction of an irreducible representation of $U(n+1)$ to $U(n)$ into irreducible representations of $U(n)$ can be easily obtained from the preceeding fundamental formula. The corresponding result has already been recalled in Section III.1.1.

Standard references for the preceding results are [36] , [37] .

This being an appendix, the authors apologize for having to omit the natural justification of the correspondence between irreducible representations and signatures (highest weights of irreducible representations of reductive Lie algebras).

NOTATION INDEX

$\mathbb{N}$, $\mathbb{Z}$, $\mathbb{R}$, $\mathbb{C}$ are respectively the set of positive integers, the set of all integers, the set of real numbers, the set of complex numbers.

H denotes a separable Hilbert space with a fixed orthonormal basis $\{e_n\}$ and scalar product $(\cdot|\cdot)$. L(H) is the algebra of all bounded linear operators on H.

For a compact space Ω , $C(\Omega)$ is the algebra of all complex continuous functions on Ω .

For a locally compact group G, M(G) is the convolution algebra of all bounded complex regular Borel measures on G and $L^1(G)$ is the ideal of absolutely continuous measures with respect to the Haar measure on G. The convolution is denoted either by "*",or simply by juxtaposition if no confusion arises. δ_g stands for the Dirac measure concentrated at the point $g \in G$.

The notations

$U(n)$, $U(\infty)$, $U_1(\infty)$, $U(H)$; $GL(n)$, $GL(\infty)$, $GL(H)$

are explained in the introduction.

<u>Note</u>. - In Chapter II the convolution is denoted by juxtaposition (see p. 63).

- For any element $x \in L$, we denote by the same symbol its canonical image in the envelopping C^*-algebra $A = \widetilde{L}$ (see p. 65).

- $n_0(t)$ is a positive integer or the symbol $+\infty$, depending on $t \in \Omega$, which indicates how many groups G_n are involved in the concrete description of $t \in \Omega$ (see p. 75-76).

<u>Note</u>. - The norm $\|\cdot\|$ $(\|P(z)\|)$ is defined on page 101.

- C_m^k stands for the binomial coefficient.

- The sign $\wedge$ stands for the exterior product (see p. 113).

Note. - In Chapter V § 2 we replace an irreducible representation $\rho_n \in \hat{U}(n)$ by its signature $(m_1,\ldots,m_n)$ in notations such as

$$\chi_{\rho_n} , d_{\rho_n} , p_{\rho_n} , B_{\rho_n} .$$

- The sign $\otimes$ stands for the tensor product.

SUBJECT INDEX

I). The following terms are used with their usual meaning, as in the monograph of J. Dixmier ([7]) :

Note. A unitary representation of a topological group is assumed to be continuous and its type is defined to be the type of the generated von Neumann algebra.

II). The following terms are defined and/or explained in the present work :

BIBLIOGRAPHY

[1]. O. BRATTELI , Inductive limits of finite dimensional C^*- algebras, Trans.Amer.Math.Soc., 171(1972), 195-234.

[2]. O. BRATTELI , Structure spaces of approximately finite dimensional C^*- algebras, J. Func. Analysis, 16(1974), 192-204.

[3]. O. BRATTELI , Structure spaces of approximately finite dimensional C^*- algebras, II, preprint.

[4]. O. BRATTELI , The center of approximately finite dimensional C^*- algebras, preprint.

[5]. J. DIXMIER , Sur les C^*- algèbres, Bull.Soc.Math. France, 88(1960), 95-112.

[6]. J. DIXMIER , Les algèbres d'opérateurs dans l'espace hilbertien (Algèbres de von Neumann), Gauthier-Villars, Paris, 1957, 1969.

[7]. J. DIXMIER , Les C^*- algèbres et leurs représentations, Gauthier-Villars, Paris, 1964, 1969.

[8]. J. DIXMIER , On some C^*- algebras considered by J. Glimm, J. Func. Analysis, 1(1967), 182-203.

[9]. S. DOPLICHER, D. KASTLER, D.W. ROBINSON , Covariance algebras in field theory and statistical mechanics, Comm.Math.Phys., 3(1966), 1 - 28.

[10]. E.G. EFFROS, F. HAHN , Locally compact transformation groups and C^*- algebras, Memoirs Amer.Math.Soc., No. 75(1967).

[11]. G.A. ELLIOTT , On lifting and extending derivations of approximately finite dimensional C^*- algebras, preprint.

[12]. L. GÅRDING, A. WIGHTMAN , Representations of the anticommutation relations, Proc.Nat.Acad.Sci. USA, 40(1954), 617-621.

[13]. J. GLIMM , On a certain class of operator algebras, Trans.Amer. Math.Soc., 95(1960), 318-340.

[14]. V. GOLODETS , Description of representations of the anticommutation relations (in Russian), Uspehi Mat. Nauk, 24:4(1969), 3-64.

[15]. V. GOLODETS , Factor representations of the anticommutation relations (in Russian), Trudy Mosk.Mat.Ob., 22(1970), 3-62.

[16]. A. GUICHARDET , Produits tensoriels infinis et représentations des relations d'anticommutations, Ann.Éc.Norm.Sup., 83(1966), 1-52.

[17]. A. GUICHARDET , Systèmes dynamiques non-commutatifs, Astérisque, No. 13/14 (1974).

[18]. P. DE LA HARPE , Classical Banach-Lie algebras and Banach-Lie groups of operators in Hilbert space, Lecture Notes in Math., No. 285(1972).

[19]. A.A. KIRILLOV , Dynamical systems, factors and group representations (in Russian), Uspehi Mat. Nauk, 22:5(1967), 67-80.

[20]. A.A. KIRILLOV , Elements of representation theory (in Russian), Nauka, Moscow, 1972.

[21]. A.A. KIRILLOV , Representations of the infinite dimensional unitary group,(in Russian), Dokl.Akad.Nauk, 212(1973), 288-290.

[22]. W. KRIEGER , On constructing non-isomorphic hyperfinite factors of type III, J. Func. Analysis, 6(1970), 97-109.

[23]. J. VON NEUMANN , On infinite direct products, Compositio Math., 6(1938), 1-77.

[24]. R.T. POWERS , Representations of uniformly hyperfinite algebras and their associated von Neumann algebras, Annals of Math., 86(1967), 138-171.

[25]. R.T. POWERS , UHF - algebras and their applications to representations of the anticommutation relations, Cargese Lectures in Phys., 4(1970), 137-168.

[26]. R.T. POWERS, E.STØRMER , Free states of the canonical anticommutation relations, Comm.Math.Phys., 16(1970), 1-33.

[27]. L. PUKÁNSZKY , Some examples of factors, Publ.Math., 4(1956), 135-156.

[28]. S. SAKAI , C^*- algebras and W^*- algebras, Springer Verlag, 1971.

[29]. I.E. SEGAL , Tensor algebras over Hilbert spaces, I, Trans.Amer. Math.Soc., 81(1956), 106-134 ; II, Annals of Math.,63(1956), 160-175.

[30]. I.E. SEGAL , The structure of a class of representations of the unitary group on a Hilbert space, Proc.Amer.Math.Soc., 8(1957) 197-203.

[31]. M. TAKESAKI , Covariant representations of C^*- algebras and their locally compact automorphism group, Acta Math., 119(1967), 273-303.

[32]. M. TAKESAKI , Tomita's theory of modular Hilbert algebras and its applications, Lecture Notes in Math., No. 128(1970).

[33]. J. TOMIYAMA , On the projections of norm one in W^*- algebras, Proc.Jap.Acad., 33(1958), 608-612.

[34]. D. VOICULESCU , Sur les représentations factorielles finies de U(∞) et autres groupes semblables, Comptes Rendus Acad. Sc. Paris, 279(1974), 945-946.

[35]. D. VOICULESCU , Représentations factorielles de type II_1 de U(∞), preprint.

[36]. H. WEYL , The classical groups. Their invariants and representations, Princeton University Press, 1939.

[37]. D.P. ZHELOBENKO , Compact Lie groups and their representations (in Russian), Nauka, Moscow, 1970.

[38]. Ș. STRĂTILĂ, D. VOICULESCU , Sur les représentations factorielles infinies de U(∞), Comptes Rendus Acad.Sc. Paris, 280(1975), Séance du 13 janvier 1975.

[39]. E. THOMA , Die unzerlegbaren, positiv-definiten Klassenfunktionen der abzählbar unendlichen, symmetrischen Gruppe, Math.Z., 85(1964), 40-61.